MW00737027

Heidegger and the Issue of Space

Alejandro A. Vallega

Heidegger and the Issue of Space

Thinking on Exilic Grounds

The Pennsylvania State University Press
University Park, Pennsylvania

Library of Congress Cataloging-in-Publication Data

Vallega, Alejandro A.
Heidegger and the issue of space : thinking on exilic grounds /
Alejandro A. Vallega.
p. cm. — (American and European philosophy)
Includes bibliographical references and index.
ISBN 0-271-02307-4 (cloth : alk. paper)
1. Heidegger, Martin, 1889–1976. Sein und Zeit.
2. Space and time.
3. Other (Philosophy)
4. Thought and thinking—Philosophy.
I. Title. II. Series.

B3279 .H48 S39 2003
111—dc21 2003005470

Printed in the United States of America
Published by The Pennsylvania State University Press,
University Park, PA 16802-1003

It is the policy of The Pennsylvania State University Press to use acid-free paper. Publications on uncoated stock satisfy the minimum requirements of American National Standard for Information Sciences—Permanence of Paper for Printed Library Material, ANSI Z39.48–1992.

A Gino y Patricia Vallega

y aquella vez fue como nunca y siempre:
vamos allí donde no espera nada
y hallamos todo lo que está esperando.

and that time was like never and always:
we go where nothing awaits us
and find all that is waiting for us.
—Pablo Neruda, *Cien Sonetos de Amor*

Contents

Preface

Exile is a term I came to know long after my exilic experience began. As I remember, my first encounter with such experience was simple and strange. I was nine years old when, together with my family, I left the world I had known as we fled from fascism and a military regime. One day we closed the door of our home in Chile, took one suitcase each (we had to leave the country under the disguise of traveling abroad for a brief vacation), and traveled by train across the Andes. The journey took a day, and we arrived at our destination at night. It is in that moment of arrival that my first exilic experience appears, and it does so as a twofold event. On the one hand, as I looked out the window I experienced a sense of total loss. My memory of this moment is that I saw nothing. It was as if I had come to a point beyond the world: no words, no visible world, not a familiar sound to receive me. This memory of loss is inseparable in intensity from a certain acceleration and excitement stemming from discovering a place, people, sounds, and words, the second aspect of my exilic experience. These sensations seemed born right there to me, in that instant, since I had never experienced or imagined the world and life that began to appear under the emerging lights of the strange city. Since then this astonishing convergence between loss and the arising of life anew has accompanied my sense of self, world, and thought. It was also this experience that led me to the term "exile," and eventually to my critical engagement with it, as I began to have a sense of the difference between exile and what I will call exilic experience and thought.

An exile is traditionally one who either by choice or force lives outside his or her country of origin. To be an exile is to be *ex patria*. This means not only being outside a city or motherland, but it also indicates separation, one's exclusion from the rights and identity given by belonging to motherland, bloodline, family, friends, language, and certain estrangement from the practices and traditions that constitute identity, as well as the grounds for making sense of life and the surrounding world. In these terms, the exile is no one and belongs nowhere. What is the place of the exile once she has abandoned her place of origin and, even under the most comfortable circumstances, has come to occupy the place of a guest? What claim under law, civil practices or everyday habits has the exile when she can no longer refer to those structures that have constituted her sense of the world, and can only at best imitate those of the host? By definition, the exile is a stranger, a foreigner, and no matter what he does, will remain foreign: once outside (ex) the place of origin there will be no return. The exile is not at home, and cannot be no matter how much he resembles the host. Indeed, once exiled, he knows that it is impossible to return. Once exiled, he was, is, and will be the foreigner, the stranger. A return only reveals how much he has changed and how much the place of origin has slipped beyond what it was, either by being still the same—in which case the one returning appears a stranger—or by having changed—in which case the one returning still finds "him-self" foreign. In either case the exile will remain the foreigner, and often a return will underscore both aspects of the slipping of the place of origin in different ways. This last point intensifies the experience of the exile by making his or her life a kind of living death. Once an exile is outside, and severed from origin, country, language, the sense of life and world that sustained existence is lost. Therefore, exile will be a living death for those who seek their identity in those unchanging and ever-present, although distant, origins.

The term "exile" figures a condemnation of all senses of life. I live as no one. I stand nowhere. Even if I take the initiative to make a life of my situation I speak and live by someone else's rules and practices (the long-lost origins or the host's ways). In this sense, as someone once said, exile is like wearing someone else's suit. This is certainly the experience of those characters in Virgil's *Aenead* who, having fled from the destruction of their city, Troy, are found weeping around a small scale version of that city. Indeed, from man's expulsion from paradise on, humanity must live an existence condemned to a veil of tears: a life that seems to find its horizons and hopes marked by the memory of a loss that will shape and judge the present in the name of that unchanging origin and the identity granted by it. This is a melancholic life for which the experience of arising from living configurations and senses of being will have never been enough, because it will have never figured a return to those unchanging origins, roots, places.

Upon close reflection and experience though, exile as defined by such a tradition does not begin to engage the exilic experience of encountering such life beyond unchanging origins in all its senses of being. The definition of exile as a being outside one's origins, country, and so forth, refers such experience *beyond* unchanging origins back to unchanging origins. The mien of the exile is the fitting exterior for the interiority of a self-determined and self-certain identity that claims to be unchanging, and therefore the measure for interpreting all experiences and senses of being. What condemns exilic experience is not its event as such, but the insistence on defining experience beyond self-identical, unchanging origins in terms of those very self-identical unchanging origins. The definition of exile is one that condemns by excluding and yet retaining sovereign power over what has been excluded. This definition never addresses the exilic experience, since it instead remains in distant judgement of it by defining it in terms of a necessary self-certain and unchanging identity.

A return to Virgil's *Aenead* gives an indication of the powerful experience figured by exilic life and thought. Aeneas is forced to leave Troy and must wander throughout the ancient world. However, his "exile" does not lead him to live as no one or to exist nowhere, always dependent on that lost city and on copying the manners of his hosts. Aeneas' experience leads to the foundation of Rome. On the one hand, this story indicates that exilic experience figures the possibility of the arising or originary event of the founding of a whole world. On the other hand, the story also points to the way the exilic experience behind the delimitation of the senses of the world can easily be covered over in the name of what arises and is given determination. Traditionally Aeneas' story and Virgil's poem find their value in light of what has been founded, in light of Rome. But one has to wonder if this allocation of the story (of the event of configuration that gives rise to such powerful world) into an economy of unchanging origins as that found in the Roman empire was not, at least in part, behind Virgil's attempt to burn the work before his death. Virgil's poem points beyond such an economy of unchanging origins, as it indicates the double character of the exilic experience that grounds Aeneas' foundation of Rome: the loss of unchanging origins (in the impossibility of a return), and the transformative effect of such experiences lead to unsuspected possibilities for the arising of senses of being. Exilic experience is not a life of nihilist negativity in the loss and absence of senses of being, of self, place, and living practices and traditions. This experience and this thought enact an event that figures not only loss but, moreover, the transformative passage of senses of being toward yet unsuspected configurations. As such, it is an experience that in its most intense manifestations refers us to the very possibility of the arising of life. Exilic experience occurs as such a dynamic event, and it is in terms of its dynamic elements (loss, transformation, possibility) that it may begin to be engaged. The questions that appear are these: How does one begin to engage exilic experiences? What do such experiences indicate about thought in its configurations and about the arising of senses of being?

It is not the personal character of these observations that is important here but what such experiences indicate. These questions open a path for engaging the exilic character of thought as well as for exploring the exilic grounds of the configurations of self, community, and the phenomenon called world. In light of these issues, I think it is not out of place to say here that this work is but a beginning step toward articulating the difficulties and possibilities opened by the engagement of exilic experience.

This book was written in many places and in a number of languages, in light of many encounters and experiences, which, in all their layers, frame and uphold the book. This particular space remains open thanks to those who have supported and shared in my work. I want to thank John Sallis, whose friendship and scholarly insight have often sustained the work in these pages; Ed Casey, for his encouragement and generous scholarship; and James Risser for his relentless support. I would also like to mention Rémi Brague and Hans-Helmuth Gander, as well as those who generously shared their insights and comments through many conversations at the Collegium Phaenomenologicum in Italy. I am grateful for the support of Helmut Kusdat, Linda Neu, Jerry Sallis, Ilya Cherkasov, and Evgenia Cherkasova. I am also grateful for the editing work of Pam Young, Daniela Vallega-Neu, Arnold Webb, and Jennifer Smith at Penn State Press. The final form of this work owes much to the close reading and comments of Dennis Schmidt. Finally, I don't know that this book would have come to be without Charles Scott, Susan Schoenbohm, and Daniela Vallega-Neu.

Introduction

Do we in our time have an answer to the question of what we really mean by the word *"being"*? Not at all. So it is fitting that we should raise anew *the question of the meaning of being.* But are we nowadays even perplexed at our inability to understand the expression "being"? Not at all. So first of all we must reawaken an understanding for the meaning of this question.
—Martin Heidegger, *Being and Time*

This question has today been forgotten—although our time considers itself progressive in again affirming "metaphysics." All the same we believe we are spared the exertion of rekindling a *gigantomachia peri tes ousias* [a Battle of Giants concerning Being].
—Martin Heidegger, *Being and Time*

The ear is receptive to conflicts only if the body loses its footing. A certain imbalance is necessary, a swaying over some abyss, for a conflict to be heard. Yet when the foreigner—the speech-denying strategist—does not utter his conflict, he in turn takes root in his own world of a rejected person whom no one is supposed to hear. The rooted one who is deaf to the conflict and the wanderer walled in by his conflict thus stand firmly, facing each other. It is a seemingly peaceful coexistence that hides the abyss: an abysmal world, the end of the world.
—Julia Kristeva, *Strangers to Ourselves*

The Alterity of the Question of Being

Heidegger's *Being and Time* calls for the reawakening of the sense of the question of being, a struggle that has been forgotten, covered over in an age rooted in the self-certainty of metaphysics and transcendental philosophies.[1] At the same time, the need for this reengagement occurs out of

1. *Being and Time,* 2 (*Sein und Zeit,* 1). Throughout this work the numeration and quotes will refer first to *Being and Time,* tr. Joan Stambaugh (Albany: SUNY Press, 1996) (hereafter *BT*), and second to *Sein und Zeit,* 2d ed. (Tübingen: Niemeyer, 1986) (hereafter *SZ*). This is an unaltered edition of the version first published in *the Jahrbuch für Philosophie und phänomenologische Forshcung,* vol. 8, ed. Edmund Husserl, and it includes Heidegger's marginal comments from his own copy. In certain passages I have chosen to use the John Macquarrie and Edward Robinson translation (San Francisco: Harper, 1962), or my own. In such cases I have noted the

an intimation of being given by the experience of withdrawal and loss of the sense of being. Furthermore, what is intimated (the question of being) is only that—an intimation that remains to be thought. Heidegger's project arises, then, from the loss of the sense of being by metaphysical/transcendental philosophy, from an intimation given in that loss, and from a possibility that remains to be engaged in order to begin to open the question of being. It is therefore not the presence of being (conceived objectively or ideally) or what can be interpreted by re-presentational thought, but a certain loss, absence, withdrawal, and uncertain futurity that figure as directions for the rekindling of the question of being.[2] In other words, the thinking of being arises not in agreement with being (that is, out of either objective entities at hand or ideal unchanging essences, and as a discourse of presence alone), but rather in an event in which the question of being remains other to the event of thought: once in withdrawal, and a second time in its futurity or its not yet being engaged as such. The question of being therefore indicates in Heidegger's project a foreign question, i.e., a question not of an identity or an unchanging essence of beings that informs and awaits thought's reach, but of the difficulty of thinking from alterity.

This alterity does not mean the abandonment of thought or being to nihilism or nonsense. It is the alterity of being that figures the matter most intimate to and distant from thought. Although the thinking of the question of being arises as a foreign question, this thought is not toward anything other than what is own-most (*eigenstes*) to thought, i.e., the question of being. Therefore, the turn to the question of being as a foreign question is not toward something other than what is of thought, but to the very event of thought. It will only be by engaging this alterity in the reengaging of the question of being that thought will come to its main question—again, the question of being. In other words, *Being and Time* calls thought to open itself to its alterity. Inasmuch as this engagement of the question of being happens in light of the loss and withdrawal of the

translation in the corresponding endnote as either M&R or mine. Throughout this work I will use "being" with a lowercase "b" to translate *Sein,* and dasein with a lowercase "d," to avoid the transcendental, metaphysical connotations suggested by their capitalization.

2. The verb and noun forms of "figure" are used throughout the book to indicate the temporal, finite, and transformative character of determinations of thought. In other words, this term is meant to at least begin to echo the alterity and exilic grounds of all conceptual determinations of beings, and in this sense it can be contrasted to the unchanging terminology of metaphysical and transcendental "ideas," "essences," "principles," "values," and their logical categories. The term "re-presentational thought" and related forms indicate that the latter thought understands itself as the repetition of what is present.

sense of being and in a move toward a thought that is yet to come, one can say that this reengagement will occur on exilic grounds, i.e., in light of the loss of ever-present, unchanging origins and as an event that in its passage enacts a transformative motion toward possibilities of being beyond already determinate configurations of beings and thought. Given the alterity and exilic character of Heidegger's project, a crucial task appears concerning *Being and Time*, as well as thought in general: if one takes up the task of reengaging the question of being, one takes up the difficulty of engaging its alterity, and the task of engaging that which is other to metaphysical thought, presence, and representational thought—that which is near to Heidegger's project only out of this withdrawal. At least three salient questions arise then: does Heidegger's *Being and Time* engage the alterity of the question of being? How will thinking engage its alterity? And how will the foreign be heard as such in its alterity?

What is crucial here is the "as such." The difficulty is that in order for the foreign to be heard as such, it must remain to a certain extent foreign. Thus, the task can not be that of subsuming or taking the foreign—the question of being—into the metaphysical and transcendental traditions of objective and ideal presence and re-presentational discourses. (There will never be an engagement of alterity if the foreign is relinquished to "the rooted one who is deaf to the conflict.") Instead, the problem for Heidegger's project, as well as for this work, is to let the question of being be heard in its alterity. Therefore, there cannot be a translation or mediation of the question into a resolution, a sublating (*aufheben*) of alterity into the rooted comfort of metaphysics or absolute knowledge (a resolution in terms of the rooted one, the other taken in in the name and under the pretense of "a seemingly peaceful coexistence that hides the abyss"). But, one could ask, is it at all possible to engage the question of being in its alterity and on exilic grounds? For Heidegger the rekindling of the question of being is never a question isolated from our age and tradition, hence other aspects of the question arise: how will one's thought bare its alterity, both in thinking in the shadow of its metaphysical and transcendental lineages and in being called to leap beyond them in order to enact a thinking in alterity—out of the loss of the sense of being, not in terms of presence, and toward a thought yet to come?

As my parenthetical remarks show thus far, Julia Kristeva's words echo the difficulty that thought encounters in being called to turn to its alterity. At the same time, she is explicit when she indicates what must occur for the foreign to begin to be heard. Kristeva writes, "The ear is receptive to conflict only if the body loses its footing. A certain imbalance is necessary, a swaying over some abyss, for a conflict to be heard." Only in a certain slipping and disruption—in the enactment of an interruption of rooted certainty—is the foreign heard as such. In terms of the rekindling of the question of being, one can say that, in order for thought to hear the question of being, an interruption, a loss of ground must occur, where thought enacts "a swaying over some abyss." It will only be in the decisive interruption (*Ent-scheidung*) of thought's claim to metaphysical and transcendental principles as the ground and root of the question of being that this very question will begin to be experienced in its alterity.

The necessity of this decisive thought brings forth a series of other questions for the reader of *Being and Time*. Does the thinking of *Being and Time* enact such slipping? In other words, is there in Heidegger's book, at least to a certain extent, a sounding out of the question of being in its alterity? And if so, how does that thought remain in suspension over some abyss that will let the question of being be heard? How does Heidegger's thought enact its alterity? Furthermore, how will one engage and follow such thought in its slipping? How will one's thought remain with that abyssal sway in a manner that lets echo the question of being in its alterity? The present work takes up the task of beginning to engage the alterity and exilic grounds of events of beings and thought in light of these questions.[3] My thesis is that Heidegger's thought in *Being and Time* performs such an interruption of thought's self-certainty, and that in this interruption a space of possibility for the engagement of the alterity and exilic character of thought and events of being is opened.

Exilic Thought

The figure of the foreigner and the task of turning thought toward its alterity are unsuspected aspects of Heidegger's project in *Being and Time*. Yet, as the present study will show, these issues belong to that

3. As explained later in the introduction, the phrase "events of beings" indicates the phenomena understood in their temporality, finitude, and concreteness, and stands in contrast to the metaphysical interpretation of such events in terms of ever-present, unchanging essences, ideas, and so forth.

thought, and as such they point to the difficult issue of "exilic thought" in Heidegger's work. One way to introduce this aspect of thought is by contrast with the exile of the question of being.

Together with metaphysics and transcendental philosophy as the ground for thought and for the interpretation of beings, one finds an economy of exclusion, dependence, and overbearing assertiveness, which enacts the silencing and forgetting—in other words, the exile—of the question of being. According to Heidegger, this interpretation of phenomena and thought is sustained by concepts such as those of unchanging, ever-present roots, origins, ideas, essences, and principles, which supplant the being of entities at hand in their temporality, i.e., their perpetual change, contingency, and mortality. Under this metaphysical duality one also finds the concepts of interiority and exteriority, limit and border, which signify the separation between specific kinds of beings. In general, although such dual systems arise out of a single focus on the presence of entities at hand, they ground the sense of beings and thought in unchanging, self-sufficient principles and their logical rules, while defining existence outside these principles as irrational and senseless, as an image of the unchanging, or as necessarily subsumed under metaphysical principles and logical rules. Finite existence in the metaphysical tradition appears as senseless negativity, and as meaningless outside of its mimetic significance, i.e., as an image of unchanging truth. I call these ways of interpreting events of beings "economies of exile" because in each case existence is understood as essentially dependent (in its configurations and functions) on unchanging principles of being. At the same time, when the phenomena cannot be defined in terms of such ever-present, unchanging essences and pure principles, it is understood as outside them, as disruptive to them, and as defective in its imperfect finitude. In this way, existence must depend on these ideal paradigms for its configurations and identities, and it must answer to pure, unchanging concepts and logical necessity if it is to have any sense at all. Beings are either inside or outside the rules and principles of reason, and the separation between unchanging Being or essence and changing beings or entities figures the exile of beings in their diversifying and unique temporal events or finitude. This occurs as the very possibility of the arising of phenomena, in its new configurations and beyond already operative determinations of beings and thought, is rejected by the rule of unchanging essences. Heidegger finds some of the strongest examples of such metaphysical interpretations in the duality of unchanging ideas and changing phenomena in Platonism, and in the metaphysics and transcendental philosophies that follow it. He finds this metaphysical dualism in the idea of an absolute teleology of existence as it is found in Aristotle and, in its historical form, in Hegel. He finds it as well in the Cartesian idea of a self-evident mind that can alone intuit and dictate the rules of being, and in the idea of transcendental principles or essences that constitute the condition for the possibility of existence, as found in Kant and Husserl.

In this metaphysical economy thought remains within a distinct delimitation. Philosophical discourse and language in general function as re-presentational tools, which have the task of making intelligible the unchanging, ever-present first principles behind appearances or beings. In other words, for the tradition, philosophy is always a giving of accounts about the unchanging Being behind beings. This mimetic force is also interpreted in terms of the production of objective and conceptual presence (rather than in terms of a *poiesis* that enacts nature's passage, for example), and in this way philosophical discourse remains occupied with its dictatorial tasks, dictating the rules and significance of existence while impervious to its own originary (*ursprünglich*) events in their essential finitude and historicity. Just as the idea of unchanging principles keeps existence in exile from events of beings, the adjudication of these rules and principles as the definitive function of thought turns thought away from its event, i.e., away from the engagement of the arising of the delimitation of the ideas, principles, rules, laws, and limits that thought is said to present, produce, and administer. In assigning to language and thought the role of dictating the principles of Being to beings, the ideality of metaphysics keeps thought from engaging the concrete finitude of its passages. The extreme forgetting of thought in the turn to metaphysical ideality and its rules is dramatically evident in the claims to a critique taken up by rationalist thought, where, even when thinking turns to look at itself, it does so in terms of assertive logical coherence and in the name of its ability to state the rules and laws that demarcate its ability for true statements and judgements. In short, in the metaphysical/transcendental tradition that sustains the history of ontology one finds figured the exile of the question of being, both in the separation of beings from Being and in rational-ideal determinations of the senses of being, as well as in the reduction of thought to the office of administrator and product provider of the senses of beings. To the reader acquainted with Heidegger it should be evident that this broad image of the exile of the question of being follows Heidegger's critique of metaphysics, while it places this critique within the question of the foreign.

This brief sketch of the exile of the question of being can serve as a contrast that lets one begin to see the issue of exilic thought. The project of *Being and Time* presents a series of specific problems for thought: beginning to think out of what is foreign to one's thought, thinking thought in its alterity, and thinking in the awareness of that thought's alterity. By the term "exilic thought" I mean to indicate such issues for thought in their density and difficulties. This thought can be directly differentiated from the history of ontology. The term indicates a thought that can not return to unchanging origins or principles in order to find its sense of being. In this sense this thought is neither determined by ideas of unchanging, ever-present origins and principles (metaphysics and transcendental philosophies), nor of ever-absent origins or roots (negative theologies), nor by a teleology that has always already determined and "taken in" thought's passage in terms of a necessary possible accomplishment or goal (either as an activity determined by actuality and potentiality, or by a dialectic leading to a fulfillment or end). In other words, "exilic" indicates a thought beyond metaphysical dualism. This very interruption of the dependence on ever-present, unchanging origins or essences also figures a transformative aspect of this thought, since it is a thought that must be engaged in its very events of passage or temporality. Furthermore, as such a temporal event, this thought also remains open to possibilities yet to come, possibilities beyond operative conceptual determinations of beings and thought. The term "exilic" indicates, then, this loss, its transformative character, and its opening toward unsuspected possibilities of being, and as such the term is not synonymous with "exile."

Exilic thought figures thought's engagement with its temporal-spatial and finite events. It indicates a thought that thinks in the awareness of its temporality, and therefore that appears as a slipping passage. As a result, it is a thought that, in the engagement of its finitude, remains aware of its alterity, hence open to the foreign in a nondeterministic manner that arises in attunement with its alterity and exilic character. In other words, "exilic" refers to a thought that remains in the open play of differences that occurs as the delimitation of identities, concepts, and ideal determinations. It is a figure of passage with neither rooted beginnings that it may reclaim without change, nor a predetermined finality that predetermines its event. This thought is exilic in that it remains in the open sway over some abyss and in that it enacts the loss of unchanging roots, principles, rules, laws, and limits. Perhaps another word that at times echoes this thought is "nomadic," a word that recalls both the ever-changing events of delimitations of being and their formal determination in the *nomoi*, the laws that give form to the projects of modern and contemporary philosophy, the sciences, and civil society. I suggest, as a last indication, that exilic thought is an indication of an experience of thought that remains in the evanescent lightness and density of its finite transformative passage, a passage that belongs to thought's alterity in the uprooting transfigurative enactment of its essential temporality or historicity (*Geschichtlichkeit*).

The Book's Project

The aim of this book is to begin to develop exilic thought through an engagement with Heidegger's attempt to recall the question of being through his analysis of the question's finitude and temporality, which includes the finitude and temporality of thought. In *Being and Time* Heidegger engages the temporality and finitude of the question of being as encountered through dasein's futurity (being-toward-death) and historicity (*Geschichtlichkeit*). The engagement of thought in terms of this futurity and historicity uncovered by Heidegger's analysis of dasein ultimately indicates the alterity of thought, and, at the same time, also points to how thought enacts an opening for the possibility of exilic thought. Three crucial aspects of the possibility for exilic thought found in *Being and Time* are the interruptions operative in the very event of the arising or beginnings of that thought; the uprooting of the metaphysical tradition through the transformative appropriation of its fundamental concepts (what Heidegger calls the deconstruction of the history of ontology); and the enactment of the alterity of thought in its temporality, which occurs as Heidegger's thought in *Being and Time* moves beyond its own conceptual limits. These three aspects of alterity are transformative in that in each case, thought's event leads beyond its already operative conceptual delimitation. They refer to exilic thought in that when Heidegger engages thought's finitude and temporality, thinking becomes a figure of passage that no longer claims its identity from unchanging concepts outside the historicity or concrete events of thought.

Both the exilic character of thought and beings and the possibility of engaging it become apparent when one looks at Heidegger's project and discourse in *Being and Time* through spatiality, and not exclusively in terms of temporality. Heidegger is clear in *Being and Time* and thereafter that spatiality is essential to the question of being. However, in the same book, Heidegger's focus on developing a discourse grounded on the temporality of the question of being leaves only a marginal place for spatiality. The question is then this: how is spatiality, an essential and inseparable aspect of the question of being, operative in the thinking of *Being and Time*? As the present study shows, from its marginal place spatiality punctuates the development or delimitation of Heidegger's discourse on temporality. Throughout Heidegger's book spatiality appears as a constant aporetic element in his discourse, which ultimately proves to be insurmountable. Spatiality appears not only explicitly and thematically, but also implicitly through various interruptions of the main discourse that point to difficulties that later will lead Heidegger to abandon his attempt in *Being and Time* to articulate spatiality in terms of the essential temporality of the question of being. Spatiality, although an issue inseparable from the question of being, ultimately remains a stranger to the main discourse of *Being and Time*: first as a marginal issue that interrupts Heidegger's discourse on temporality, and later as an insurmountable difficulty for the way the question of being is raised in that book. The aporetic moments and the alterity of the spatiality of the question of being serve as a constant trace of the delimitation of Heidegger's thought in *Being and Time*; and the event of this thought occurs as a delimiting play between Heidegger's conceptual determination of the question of being (in a discourse of temporality) and what remains always beyond this thought and ultimately calls for the thought's transformation (spatiality). In other words, spatiality appears as an exilic figure in the discourse on temporality, and at the same time, when engaged, it indicates issues beyond that discourse. In light of the character of spatiality as a figure of alterity and exilic thought, the alterity of the question of being in *Being and Time* and the exilic character of that thought are made apparent by remaining with Heidegger's struggle for the question of being in his difficulties with spatiality—struggle and difficulties that, to echo Kristeva's words, mark the losing of footing of Heidegger's discourse of temporality, and the opening toward the alterity and exilic character of thought given in that fall.

From this brief introduction it should be clear that this project is not merely a commentary on *Being and Time*. Its task is not that of rehearsing what Heidegger says and his intended project. Rather, it is an engagement with issues essential to his thought—namely, the alterity and exilic character of thought—and, along these lines, it is a work on the way toward the development of what I have called exilic thought. Because of the particular task of this work I have followed Heidegger's thought not only with attentiveness to what he says but also, more importantly for this project, with attention to the performative or enacting aspects of his thought. This means that I have not only traced the structure of the book in its *phenomenological* unfolding of the question of being and closely interpreted Heidegger's dasein analysis, but that I have also focused on listening for and engaging thematically the interruptions, the moments of silent reticence, and the hesitations that mark Heidegger's engagement with spatiality in *Being and Time*. Indeed, it is precisely in these moments of Heidegger's discourse that the alterity of his thought begins to be heard.

Before going on to the outline of the book some further observations must be made concerning the role of spatiality in this work. Spatiality appears in this work as what I call a figure of alterity. It can neither be given full determination through Heidegger's discourse on temporality as origin of all beings, nor in terms of traditional metaphysical and transcendental thought. Therefore, spatiality will remain beyond presence, although essential to the presencing of beings. At the same time, this issue of spatiality points to the concrete finitude of events of beings. The term "spatiality" does not indicate an abstract or purely conceptual aspect of beings or thought. Spatiality recalls thought to issues of physicality, embodiments, and communal and political questions. In all our senses of being, in all that is said to be and at the articulate limit of what can be said, one finds a certain sense of spatiality. Indeed, even in the attempt to speak of not being, one experiences a vertigo open to a seemingly infinite space. Life, bodies, communities, things, words, numbers, thoughts, feelings and states of mind, our imagination—all these occur not only temporally but also spatially, if one understands spatiality not just as *res extensa* or its ideal analogies (for example, the transcendental space of reason in Kant or Husserl's universal space). At the same time, because these senses of spatiality are never reducible to presence, they are never graspable by accounts or measurements (as happens in mathematical physics), nor are they explained by any thinking fixed on the infinite production of things, quantities, feelings, or images, as is the case in our age of technological production. Strangely, spatiality appear with all beings and yet remains other, as if withdrawn in letting beings appear in all their senses "in space." Spatiality seems always to be at our fingertips and yet remains immaterial and uncanny. It is this strange sense of the being of spaces that gives direction to the following work, direction not in the sense of something already ahead waiting to be thought, but in the sense of a difficulty essential to any philosophical attempt to engage events of beings in their concrete finitude and possibilities.

Three other observations concerning terminology will help in the reading of the book. In it I speak of "events of beings" and "occurrences of beings," and relate these terms to "beings," "the question of being," and "being." The first two terms indicate the temporality of the question of being in its concrete finitude, in its events. As I indicate in Chapter 2, in *Being and Time* as a whole, in its three parts, Heidegger's project is neither "Being" nor "beings" understood as entities at hand, but ultimately to think beings out of their essential temporality. This is why I use "events of beings" as analogous to "being," where the "of" is not meant as a genitive that refers the question of being to beings (understood as entities), nor vice versa (two ways of still reinscribing Heidegger's thought into the metaphysical dualism of objective and ideal presence). "Events of beings" indicates the temporality and finitude of the question of being and of beings in their unique, concrete, and ephemeral events. A second term that requires clarification is "objective and ideal presence." This phrase refers to the metaphysical and transcendental interpretation of events of beings in terms of a double presence: as the changing material presence of entities at hand, and as the unchanging, ever-present ideas and principles that order, direct, determine, and make possible all phenomena. The order of the terms, first "objective" and then "ideal," is purposeful, and it points to Heidegger's understanding of Western metaphysics and transcendental philosophy as arising out of a certain interpretation of events of beings in terms of the logical necessities of entities present at hand. The third term I must specify is "space." When in quotation marks, it refers to the interpretation of spatiality in terms of the metaphysical dualism of objective and ideal presence, i.e., as the space of entities at hand and as the ideal or transcendental space that is the possibility for such empirical experiences. When the term is used without quotation marks it refers to senses of spatiality that are not understood in such metaphysical terms (for example, "the space opened by Heidegger's thought"). This sense of space indicates the spatiality of events of beings and thought in their presencing or temporality, and therefore figures events not grasped in terms of presence.

Brief Outline of the Book

This work is divided into three parts: "Themes" (Chapters 1 and 2), "Scherzi" (Chapters 3–6) and "Fugues" (Chapter 7).[4] As the title of the first section indicates, the first two chapters introduce the main themes of the book: spatiality, alterity, exilic thought, and the difficulty of finding a language that engages these aspects of thought and beings. Chapter 1 traces the problem of spatiality in the question of being back to Plato's *Timaeus* and Aristotle's *Physics*. The chapter shows that for both ancient philosophers the issue of spatiality is essential to the question of being, and that in their works it appears as a figure of the alterity and exilic grounds of thought and beings. These issues are discussed in light of the ancients' interpretation of the philosophical logos as a mimetic or presentational discourse and a tool for presenting the unchanging, ever-present principles behind nature's constant change. In light of Heidegger's introduction to his project in *Being and Time* as a rekindling of the question of being, and as a question given in the thought of Plato and Aristotle, these observations introduce issues essential to Heidegger's book. Chapter 2 considers spatiality in *Being and Time* and shows how it figures the alterity of thought and events of beings. This discussion occurs in relation to Heidegger's project of an "apophantic logos," i.e., philosophical thought as a transformative hermeneutics, which in its engagements of the question of being aims to go well beyond Plato's and Aristotle's representational approaches to the philosophical logos. My focus on this difference in thought and discourse indicates the exilic character of Heidegger's project.

The second part of the work is titled "Scherzi," and it is a series of four encounters with the figure of spatiality in *Being and Time*. Each of the

4. The musical terms explicitly refer to the function and motion of each part, while at the same time, indicating an encounter with Heidegger's thought guided both by what is said and by attention to issues of attunement, listening, and silence. These are tacit aspects of thought often brought to our awareness by music and musical terminology.

four chapters indicates ways in which spatiality figures the essential alterity and exilic grounds of Heidegger's thought in that book. These discussions arise in light of the difficulty of a logos or thought that attempts to engage the alterity and the exilic grounds of events of beings, as well as its own event, in a way other than through the representation of unchanging origins, principles, and teleologies. Chapter 3 discusses the "twisting free" of the figure of spatiality from its traditional interpretation in terms of objective and ideal presence, a release that occurs through Heidegger's critique in *Being and Time* of Descartes' ontology of the world. The chapter closes with a reflection on thought's exilic grounds as enacted by Heidegger's critique of Descartes, insofar as Heidegger's thought cannot be reinscribed into the transcendental-metaphysical tradition after the critique, but at the same time is not yet grounded on dasein's temporality and finitude.[5] Chapter 4 engages Heidegger's transcendental articulation of spatiality in *Being and Time*. This articulation ultimately appears as a "failure," and is rethought as the enactment of the alterity and transformative overcoming essential to Heidegger's thought, since Heidegger's temporal analysis leads beyond its very attempt to situate the origin of events of beings in temporality. Chapter 5 takes up Heidegger's discourse on spatiality in *Being and Time* and focuses on his attempt to reappropriate the traditional language of "space" through a transformative appropriation of it in his discourse on both dasein's spatiality and spatiality in general. The translation is then discussed as the enactment of the revealing-concealing structure or the alterity essential to Heidegger's philosophical logos in its temporality and finitude or ephemeral passages. The last chapter of this section, Chapter 6 reengages the previous three chapters' discussions of alterity, exilic grounds, and philosophical logos in *Being and Time* by placing them in relation to Heidegger's analysis of dasein's being-toward-death, or ekstatical being and ekstatical spatiality. The discussion shows that dasein's being-toward-death figures the alterity and exilic grounds of thought and events of beings. The chapter then turns to *Being and Time* and looks at it as a thinking on and out of exilic grounds. In light of Heidegger's analysis of dasein's ekstatical being, and through the various passages of

5. The doubling of temporality and finitude and vice-versa is meant to indicate the inherent inseparability of concreteness and temporality in Heidegger's thought. At the same time, this doubling should at least recall the difficulty that it opens in Heidegger's work and in contemporary thought.

thought discussed in previous chapters, Heidegger's discourse in *Being and Time* appears as the enactment of its transformative event, an event that in its passage is marked by the figure of spatiality as a deconstructive element in Heidegger's discourse on temporality.

The last section of this work, "Fugues," is a long chapter engaging variations on the issues of spatiality, alterity, and exilic grounds in Heidegger's later works. The discussions follow various figures of spatiality and direct attention to how, through these figures, Heidegger engages the concreteness and alterity of events of beings on exilic grounds.

PART ONE

Themes

1

Transgressions

Recalling the Alterity of Beings in Plato and Aristotle

Introduction

There are many points that make up the beginning of *Being and Time*, but in the first pages of the book Heidegger orients his attempt to regain a sense of the question of being by recalling the thought of Plato and Aristotle. The book's first sentence is not in German but in ancient Greek, a quote from Plato's *Sophist*: "delon gar hos humeis men tauta (ti pote boulesthe semainein hopotan on phthengese) palai gignoskete, hemeis de pro tou men oometha nun d'eporekamen."[1] What is one to make of such a beginning? Why recall Plato and Aristotle, the philosophers who stand at the beginning of metaphysics? What is it that these thinkers bring to Heidegger's project? Without presupposing the character of Heidegger's relationship to Plato and Aristotle, or, for that matter,

1. *BT,* 1 (*SZ,* 1).

our own relationship to them, these questions are enormously difficult. Is it self-evident that the question of being can be rethought in relation to Plato and Aristotle's thought, i.e., outside of its metaphysical beginnings and somehow in a motion beyond such beginnings? And if so, how does this possibility affect the contributions that these thinkers make to Heidegger's project?

Another quote in *Being and Time* from the *Sophist* begins to indicate the importance of Plato and Aristotle for Heidegger:

> The question has today been forgotten—although our time considers itself progressive in again affirming "metaphysics." All the same we believe that we are spared the exertion of rekindling a γιγαντομαχια περι της ουσιας [Battle of Giants concerning Being]. But the question touched upon here is hardly an arbitrary one. It sustained the avid research of Plato and Aristotle but from then on ceased to be heard as *a thematic question of actual investigation*.[2]

While in the age of metaphysics the sense of the question of being has been lost, the ancients engaged in a battle of giants concerning being. The intensity of Plato's and Aristotle's thought was sustained by the question of being as "a thematic question of actual investigation." In these philosophers Heidegger finds thought guided by the question of being. In *Being and Time*, he seeks the engagement with this question in its utmost intensity. But how do Plato and Aristotle open a path for Heidegger's struggle for the question of being?

The quote from the *Sophist* that opens *Being and Time* indicates neither a conceptual nor a methodological ground for Heidegger's project. Instead, it introduces the project of *Being and Time* by way of an interruption: "For manifestly you have long been aware of what you mean when you use the expression 'being' [*seiend*]. We, however, who used to think we understood it, have now become perplexed."[3] This interruption concerns a suspension of thought's self-certainty concerning the question of being. What is suspended in the Greek is a way of interpreting the meaning (*boulesthe semainein*) of being (*on*). The call for this suspension occurs out of a certain "perplexity" or "difficulty" with the understanding of the meaning of being. The word that indicates this is *eporekamen*,

2. *BT*, 2, italics in original (*SZ*, 2).
3. *BT*, 1 (*SZ*, 1).

a word that bespeaks neither the difficulty of a rational puzzle nor the resolvable perplexity of reason alone, but a certain pause, a break, an interruption. The verb *aporeo* means "to be at a loss," "to be left without means or resources to go on." This verb can be traced back to *aporos*, a noun meaning "without passage," "an impassable point," "trackless," and "pathless."[4] Indeed, it is the recalling of such an interruption of the philosophical logos that opens *Being and Time*. At the same time, while this break in the smooth continuity of thought's discourse figures the forgetting of being, it also figures the possibility of Heidegger's project. On the one hand, Plato's and Aristotle's thought inaugurates the metaphysical tradition that, according to Heidegger, has left being to oblivion. On the other hand, the ancients' thought will call forth the question of the meaning of being from the withdrawal of being, in the loss of the sense of being enacted by their very thought. Heidegger's statement that we no longer have a sense of the meaning of being comes in light of the withdrawal of being and in our sense of the loss of the sense of being, and is at the apogee of metaphysics.[5] The intimation of the question of being that opens *Being and Time* is given in an "abyssal" moment when thinking breaks down, and in this falling is called to being in an intimation given by the absence of a sense of being.

The interruption from the *Sophist* indicates a leap toward the question of being. The break in the philosophical logos, the point when the philosopher no longer has words, the moment of silence and withdrawal is the occasion for the arising of thought. But Heidegger is quick to fill the aporetic opening of the interruption with a specific question. Immediately after the quote from the *Sophist* and his translation, Heidegger translates this interruption into the specific aim and task of *Being and Time*:

> Do we in our time have an answer to the question of what we really mean by the word "being"? Not at all. So it is fitting that we should raise anew the question of the meaning of Being [*die*

4. Liddell and Scott, *A Greek-English Lexicon* (Oxford: Clarendon Press, 1989).

5. As is well known, for Heidegger the history of metaphysics extends from Plato and Aristotle to Nietzsche. It is Nietzsche who, in Heidegger's thinking, marks the last breath of metaphysics, Nietzsche who announces the death of God and all ideal principles, and at the same time Nietzsche who gives the last metaphysical accounts of beings. "The five main rubrics we have mentioned—'nihilism,' 'revaluation of the values hitherto,' 'will to power,' 'eternal recurrence of the same,' and 'overman'—each portray Nietzsche's metaphysics from just *one* perspective, although in each case it is a perspective that defines the whole." Heidegger, *Nietzsche*, vol. 4, ed. David Farrell Krell (San Francisco: Harper, 1987): 9.

> *Frage nach dem Sinn von Sein erneut zu stellen*]. But are we nowadays even perplexed at our inability to understand the expression "being"? Not at all. So first of all we must reawaken [*wieder wecken*] an understanding for the meaning of this question. Our aim in the following treatise is to work out the question of the meaning of *being* and to do so concretely. Our provisional aim is the interpretation of *time* as the possible horizon for any understanding whatsoever of being.[6]

For Heidegger the battle of giants concerning being will be rekindled by thinking the essential temporality and finitude of the question of being. This beginning figures a leap from the interruption of thought to the introduction of a project for thinking beings out of their essential temporality and finitude.

Heidegger's powerful leap toward temporality as the essential ground of the question of being, which in its temporal event is the grounding of the occurrences of beings, overshadows the violent interruption of thought figured in the quote from the *Sophist* that opens *Being and Time*. In that passage the philosophical logos has reached a breaking point, marked by the absence of means or a path that will ground or direct thought. Indeed, the philosophical logos has reached its limit and has become speechless. What one faces here is not merely the well-known problem of the forgetting of being by metaphysics that Heidegger emphasizes in his introduction to *Being and Time*, as well as in many other texts and lectures.[7] In its basic force, the interruption in the *Sophist* that opens *Being and Time* indicates the way the question of being begins to be engaged in Heidegger's book, i.e., out of a thought that arises in light of an event of interruption, and in the loss of the sense of being and the call for thinking found in that loss. This possibility of the question of being appears only as something foreign, strange to thought. Being is given to thought in the collapse of thought's determinations of being and in the absence of a sense of being. Thus, thought is not a matter of returning to a lost or hidden sense of being, but a matter of engaging the occurrences of beings out of a certain alterity (interruption, loss) operative in the configuration of such events of being. This alterity of events of beings and thought figured in the question of being is echoed by the passage

6. *BT,* 1 (*SZ,* 1).

7. If metaphysics and the Greeks marked only the forgetting of the sense of being, the question of being would not be an issue for *Being and Time.*

from the *Sophist* as it opens and provides leeway for Heidegger's thought. One finds the echo of this alterity in the opening of *Being and Time*, an opening in a foreign tongue, an opening that, according to Heidegger, must remain in Greek as a figure of the untranslatable experience of the philosophical logos.[8]

At the same time, in his leap to the issue of temporality, Heidegger seems to forget the alterity of thought, as this is covered over by his thinking in the name of temporality. Heidegger's focused engagement of the question of being in terms of temporality/finitude also obscures the primary role of spatiality in the question of being. When one looks closely at Plato and Aristotle it is clear that for them the issue of spatiality is inseparable from the occurrences of beings. This is evident in Plato's well-known discussion of *chora* in his *Timaeus*,[9] as well as in Aristotle's rethinking of the *Timaeus* in his *Physics*.[10] This inseparability of the question of being from the issue of spatiality is echoed throughout the philosophical tradition. Generally, before Heidegger's *Being and Time*, the issue of the spatiality of being is associated with "space," and is interpreted in terms of metaphysical and transcendental traditions. Throughout the tradition "space" has had a double meaning, indicating both the objective space of particular things and the general space "wherein" things "take place." Its being is understood in many ways under this twofold interpretation: as the "space of nature," as a metaphysical category, as a substance, or as part of a transcendental consciousness. We find examples of the latter, which are directly pertinent to Heidegger's thought, in Kant's *First Critique*, in the work of Husserl,[11] and in the development of Husserl's work, particularly with respect to the issue of spatiality as approached by Heidegger's colleague and contemporary Oskar Becker.[12]

The issue of the spatiality of being and beings in Aristotle and Plato is not extraneous to *Being and Time*. Aristotle's text on this theme (in the

8. One must ask, then, to what extent is Heidegger's thought in its relationship with the Greeks sustained by this sense of alterity?

9. Plato, *Timaeus*, tr. R. G. Bury (Cambridge, Mass.: Harvard University Press, 1989).

10. Aristotle, *Physics*, vols. 1–4, tr. Wicksteed and Cornford (Cambridge, Mass.: Harvard University Press, 1989).

11. "Foundational Investigations of the Phenomenological Origin of the Spatiality of Nature," tr. Fred Kersten, and "The Origin of Geometry," tr. David Carr, in *Husserl: Shorter Works* (Notre Dame: University of Notre Dame Press, 1981).

12. Oskar Becker, *Beiträge zur phänomenologischen Begründung der Geometrie und ihrer physikalischen Anwendung*, 2d ed. (Tübingen: Niemeyer, 1973).

Physics), is explicitly addressed by Heidegger from the start of Part I, §12, of *Being and Time*.[13] The analysis of dasein's being-in-the-world (its ontological structure) is introduced through an "orienting" discussion of what "being-in" (*In-Sein*) means. Heidegger's passage differentiates between the spatiality of things at hand, which are understood as being "in" space (here space is understood as a "vessel"), and "being-in" as the particular way of dasein (being-t/here).[14] This passage from *Being and Time* paraphrases Aristotle's discussion of the same issue in Book IV of the *Physics*, "on the many senses things are said to be in another."[15] Aristotle's discussion focuses on the problem of "place" (topos) and on the spatiality of entities at hand,[16] and ultimately characterizes place as a containing vessel. Both the inseparability of spatiality from the question of being for Plato and Aristotle and the presence of this issue in *Being and Time* indicate that when Heidegger takes up the task of rekindling the battle for the sense of being, he also inherits the difficult issue of the spatiality of events of beings. But what is it that Heidegger inherits here? What does the issue of spatiality bring to the question of being?

As we shall see, the turn from Heidegger to the thought of Plato and Aristotle shows that in the engagements of the question of being, alterity and spatiality play essential roles. The relevance of these two issues for the question of being is exposed below through the analysis of *chora* in Plato's *Timaeus* and of *topos* in Aristotle's *Physics*. These explorations of the spatiality of events of beings also indicate the alterity of the question of being, and, in this light, indicate the limits of the philosophical logos with regard to its assertive, representational, and logical power and determination.[17] The uncovering of spatiality, alterity, and the limit of repre-

13. Although the text of the *Physics* is directly evoked by Heidegger's paragraph, Aristotle's thematic treatment of "being-in" is only implicitly, silently introduced in Heidegger's discussion. It is only by suspension that Aristotle's understanding of "place" is present in Heidegger's text. Heidegger's interpretation of Aristotle emphasizes motion (temporality) over place in the question of being, because of the narrow characterization of the Being of beings (and their spatiality) in Aristotle.

14. Throughout this work I keep the German *dasein*, however, when helpful as an indication of dasein's finitude, uniqueness, concreteness, and place, I also include the English translation "being-t/here."

15. "Meta de tauta lepteon posachos allo en allo legetai." Aristotle, *Physics,* IV: 3, 210a14–15.

16. *Physics,* IV: 1–8, 208a27–30.

17. In my reading of the *Timaeus* and in my rereading of Aristotle's *Physics,* I have followed the readings of Plato by John Sallis and Jacob Klein. These two approaches differ in conclusion, but share a kind of uncommon clarity and awareness with respect to the way the *Dialogues* can be engaged. See John Sallis, "Timaeus' Discourse on the *Chora,*" lecture manuscript: Boston

sentational and logical discourse in Plato and Aristotle will ultimately provide thematic points for engaging Heidegger's analysis of the question of being in *Being and Time* in terms of spatiality and alterity. These are elemental issues Heidegger seems to cover over in his leap from the interruption in the *Sophist* that opens *Being and Time* toward the question of being as a question of essential temporality.

Chora: Plato's Figure of Alterity

In Plato's *Timaeus*, one of the main speakers and the work's namesake tells a "likely story" (*logon ton eikota*)[18] that attempts to show how all beings[19]—the cosmos (*kosmos*)—came into being.[20] A close look at this story reveals interruption, difference, and alterity to be essential in the arising of events of beings. These issues are figured in Timaeus' story by *chora*. Before taking up this term, and in order to engage Timaeus' discourse, a few remarks about the story's context and aim are necessary.

What does one hear in the words "cosmos" or "likely story"? These terms point to a specific way of engaging events of beings, and as such point to a conception of the world and language that, although foundational for today's thought, is at the same time foreign to today's experience of world and facts, and to most notions of what is meant by being. This is not the place for comparisons and contrasts between ancients and moderns, but it will be helpful at least to keep in mind that there are crucial differences between the two and to identify some of the aspects of Timaeus' task that can easily be overshadowed or taken for granted by today's rationalist and technological interpretations of world and language. This is a danger particularly concerning today's ways of interpreting the cosmos as a "universe"—as unified and given form by a single

University, 1995; *Being and Logos* (Indianapolis: Indiana University Press, 1996); and *Chorology: On Beginning in Plato's 'Timaeus'* (Indianapolis: Indiana University Press, 1999); Jacob Klein, *A Commentary on Plato's Meno* (Chicago: University of Chicago Press).

18. *Timaeus* 29d; cf. 48d. *Eikota* comes from the verb *eiko,* which means "to resemble" or "to be like." The substantive form is *eikon,* a "likeness," "image," or "portrait."

19. "Hemas de tous peri tou pantos logous poieisthai pei mellontas, he gegonen e kai agenes estin." *Timaeus,* 27c.

20. *Timaeus,* 28b. The discussion concerns what is "created" (*gegonen*). See also *Timaeus,* 27c–91c. For a thorough discussion of the question of the relationship between generation (*techne* and *poiein*), logos, and *muthos* in the *Timaeus,* see Sallis, *Chorology.*

order (*uni-versus*)—instead of engaging it as an event of differences (cosmos refers the looks of particular and diversified beings in their being together). This difficulty is operative throughout today's life with respect to the interpretation of the universe as objective and as ruled by mathematical principles, and therefore as subject to determination and manipulation by rational-calculating discourses and their techno-scientific projects. This modern interpretation of occurrences of beings includes the traditional interpretations of language as a descriptive-calculating tool concerned with giving accounts of beings both in terms of presence and in the form of assertive logical arguments and descriptions by reference or analogy.[21]

In the *Timaeus* there is no word for universe as we now understand it. The cosmos is a gathering of diversified events of beings.[22] As the term indicates, cosmos involves the occurrences of beings in their diversified looks. One might call this cosmos "concrete," if this word is taken to mean the appearing of beings in their diversity and in the inseparable physicality and conceptuality of their events.[23] It is in light of this concrete experiencing of the occurrences of beings that Timaeus will tell his story. Thus, to use a figure from Heidegger's thought in the 1930s, Timaeus' likely story concerns the beings of earth, sky, mortals, and gods, all that is experienced in the sensible-intelligible world.[24]

This concrete experience is not reducible to the objectifying and abstracting of today's pragmatist and rationalist ways of approaching experiences of beings, i.e., experience in Timaeus' sense is neither a fact nor an abstract idea to be analyzed and then explained away. Taken in their concrete sense, the appearances mark and expose the very limits of their events. From its beginnings and throughout its tradition, Greek

21. For an introduction to the problem of ancients and moderns, see Jacob Klein's *Greek Mathematical Thought and The Origin of Algebra* (New York: Dover, 1992) and *Lectures and Essays* (Annapolis, Md.: St. John's College Press).

22. The Greek word that is often equated with universe is *to pan,* which literally means "the all" and therefore is not exclusive to a single order or principle of being.

23. This sense of events of beings is echoed in Plato's "idea," the form or shape of beings, a concept that entangles sensible and conceptual knowledge in the intelligibility of the events of beings.

24. See Heidegger's "Origin of the Work of Art," in *Basic Writings* (San Francisco: Harper, 1993); "Building, Dwelling, Thinking," in *Vorträge und Aufsätze* (Stuttgart: Neske, 1985); and *Contributions to Philosophy* (Indianapolis: Indiana University Press, 1999). The two terms of "sensible-intelligible" refer to the metaphysical dualism that defines the world in terms of two kinds of presence, the presence of the sensible and changing and that of the intelligible and unchanging. Both kinds are grounded on presence.

thought engages all sensible-intelligible appearances by taking them in the possibility and question of their sense of being.[25] Greek thought takes up sky, earth, mortals, and the divine, and attempts to question them in their being, in their appearing. This is the way of being in the world that is evident when one looks at the Greek experiences in epic, tragedy, and philosophical discourse. Some of its more dramatic examples are found in Homer's Odysseus, who must travel beyond the limits of the sensible-intelligible world to reclaim his identity; in Sophocles' Oedipus, whose task is to know even at the expense of all that gives sense to his being a king, citizen, and to the very order of the city; in Heraclitus, who states that being loves to hide and that one must battle for truth as if standing on the walls of the city; and in Plato's Socrates, who speaks both in the *Republic* and the *Phaedo* at the edge of the underworld, at the edge of the sensible and the intelligible. All of these figures experience being at the limits of already operative senses of being as they relentlessly question these limits. In other words, the Greek logos travels to the limits of the sensible-intelligible order or world in order to ask about the sense of the occurrences of beings. At this limit all senses of beings are in play. On the one hand, they remain to be thought; on the other, they are engaged as being at the mercy of fate and disaster, as well as human action and thought.[26] In such experience at the limits of all sense of the occurrences of beings, nothing is given; and yet thought undergoes the delimitations of all senses of beings in their concrete events.

As Timaeus himself indicates when he calls attention to his likely story, these dangerous journeys to the limits of the senses of being always involve a logos. These *logoi* are the speeches of mortals, which spread like nets in an open sea in their engagements with events of beings. The logos marks paths to the limits and possibilities of the senses of the occurrences of beings. At the same time, the logos only occurs as a return with nets full of intimations and echoes of events of beings. This vast net of the logos spreads throughout all appearances of sensible and intelligible being. But again, the logos is neither a matter of recording facts nor a set of calculating and manipulative tools. The logos marks the appearing of

25. Heidegger indicates this in his discussions of phenomenology in terms of *logos* and *phainomena* in *BT,* 30–34 (*SZ,* 34–39), also when he asks "[i]s it a matter of chance that the Greeks, who at the same time had eyes to see, determined the essence of human beings as *zoon logon echon*?" *BT,* 154 (*SZ,* 165).

26. See Charles Scott, "*Adikia* and Catastrophe: Heidegger's 'Anaximander Fragment,'" in *Heidegger Studies* 10 (1994): 127–42.

all beings at the limits of their sense and therefore also in their arising to their possibilities. The famous chorus from Sophocles' *Antigone* sounds out the widespread and limited power of the human logos:

> Many are the wonders, none
> Is more wonderful than what is man. . . .
> A cunning fellow is man. His contrivances
> Make him master of beasts of the field
> And those that move in the mountains. . . .
> And speech and windswift thought
> And the tempers that go with city living
> He has taught himself, and how to avoid
> The sharp frost, when lodging is cold
> Under the open sky and pelting strokes of the rain.
> He has a way against everything,
> And he faces nothing that is to come
> Without contrivance.
> Only against death
> Can he call no means of escape . . .[27]

As the last lines indicate, the logos itself encounters its limits in its events and determinations of beings. The logos always echoes its finitude, and hence a certain indeterminacy essential to its events. As such, the logos itself is always in question and at play in its many returns, since its very events seek to articulate the question of the coming to pass of events of beings.

It is a dangerous and difficult journey to the limits of the sensible and intelligible senses of events of beings that Timaeus undertakes as he weaves his likely story, always aware of the danger of losing all senses of beings in the engagement. Just as Odysseus must transgress the limits of the sensible and intelligible world in order to reclaim his identity, and just as Socrates begins to speak at the edge of the underworld, Timaeus speaks in a way that invites the reader to engage all that is, the sensible-intelligible cosmos at its limit, at the limit of what can be said to be. Only as one engages in the dangerous and concrete journey of the logos does Timaeus' story begin to resonate in its intensity and difficulty.

27. Sophocles, *Antigone* (Chicago: University of Chicago Press, 1991), lines 368–95.

Timaeus' Likely Story

When one looks at Timaeus' likely story it is evident that his account is made up of at least two stories. Timaeus begins by giving an account of the origin of the cosmos according to what can be gathered through reason (*ta dia nou*). He presents two forms, "being," or "what is always and never becoming" (*ti to on aei, genesis de ouk echon*), and "becoming," or "what is always becoming and never is" (*kai ti to gignomenon men aei, on de oudepote*). Then he generates the cosmos—all things sensible, tangible, and with bodies (*gegonen: horatos gar haptos te esti kai soma echo, panta de ta toiauta aistheta*)—as a copy of being in becoming (*ton kosmon eikona tinos enai*). This image is fashioned out of the four elements (air, water, fire, and earth) and in accordance with one self-identical cause only reachable by reason (rather than the senses).[28] The story is then interrupted, since it has reached a certain limit.

A certain incompleteness in the first account leads Timaeus to a new and different beginning. In his first account Timaeus has taken the four elements for granted, as already given.[29] To avoid this assumption, he must then make a second and perhaps more originary beginning; he begins the second account by giving an account of the generation of the four elements.[30] This time, in order to give a complete account of the coming into being of the cosmos, Timaeus adds to the two forms of his first account a third. In addition to being and becoming, Timaeus introduces *chora*.[31] Even when first introduced this figure marks the strangeness of Timaeus' story. *Chora* is different from the other two forms in that it appears outside of the limit of what may be gathered by reason. Timaeus' second beginning occurs after having reached the limits of his account according to reason. In its difference, *chora* traces a path of "necessity," and is introduced as an "errant cause," rather than following the order of reason to a self-identical cause of the cosmos.[32] Furthermore,

28. See *Timaeus,* 27c–48; 28; 28b; 29a; 29b; 31b; 47e; cf. 32c.

29. *Timaeus,* 48b.

30. Throughout the *Timaeus* one finds the play of the originary generation of the cosmos and Timaeus' account as a demiurgic activity. See Sallis, *Chorology.*

31. *Timaeus,* 48e–49.

32. We are speaking of a cause that in its motion does not draw a perfect circular path, but the elliptical and oscillating one of celestial bodies. The word "errant" in the Greek is *plane,* hence the use of the term "planet" to speak of the celestial bodies, where the term indicates the irregular, wandering motion characteristic of celestial bodies. In *The Fate of Place* (Berkeley and Los Angeles: University of California Press, 1997) Edward Casey makes the good point that "Cosmos"

as such, *chora* is a form outside the metaphysical dualism between beings as entities and being as the ever-present, unchanging first principles and causes behind entities. Thus, although called a "kind" or "form," *chora* is not a kind understandable in metaphysical terms. Rather, *chora* indicates a certain strangeness, or at least intimates a certain aspect of events of beings that slips from or escapes reason in its interpretation of the world as ideal and objective presence.

In his search for a complete account Timaeus has now given not one unified account but two different accounts with two different and, as we will see, irreducible beginnings for the cosmos.[33] The first account generates the cosmos according to the laws of *dianoia*, and it goes as far as it can go through reason (*ta dia nou*).[34] The second account begins from this limit, and goes outside, beyond the dianoetically conceived cosmos. In short, the account of the origin of beings is now a double account held in its difference by two irreconcilable beginnings, irreconcilable because what stands between them, in Timaeus' account, is the interruption of rational discourse and order, a loss of a single unifying principle or order under which the difference may be subsumed. This irreducible difference is evident in Timaeus' own description of how the two beginnings are mixed together in the cosmos' coming into being: "For in truth, this Cosmos in its origin was generated as a compound, from the combination of necessity and reason. And in as much as reason was controlling [*archontos*] necessity by persuading [*peithein*] her to conduct to the best end the most part of the things coming into existence."[35] The two causes are held

---

does not mean the same as "Universe." The idea of a unified, ever-hovering, and receiving "space" does not necessarily agree with the "Cosmos," which is a gathering of diverse living elements. "'Uni-verse,' *universum* in its original Latin form, means turning around one totalized whole. The universe is the passionate single aim of Roman conquest, Christian conversion, early modern physics, and Kantian epistemology. In contrast, 'Cosmos' implies the particularity of place. Taken as a collective term, it signifies the ingrediency of places in discrete place-worlds. (The Greek language has no word for 'universe,' instead, it speaks of *to pan,* 'all that is,' 'the All')" (78). Cf. Remi Brague's interpretation of Aristotle's idea of Cosmos, or *uranos,* in terms of world, in *Aristote et la Question du Monde* (Paris: Presses Universitaires de France, 1988).

33. It is worthy of attention that the incompleteness of the first account leads to the second, i.e., the second beginning occurs out of a certain incompleteness, and not out of any self-certainty or secured first principle that will guarantee the completeness of the second beginning.

34. "The foregoing part of our discourse, save for a small portion, has been an exposition of the operations of reason [*ta dia nou*]; but we must also furnish an account of what comes into existence through necessity." *Timaeus,* 48a.

35. *Timaeus,* 48a.

together by reason's "persuasion" (*peithein*),[36] an expression that elsewhere in the dialogue indicates a form of compulsion.[37] *Peithein* means "to prevail upon" or "to win over," even by misleading words or a bribe. We are speaking of bringing the two beginnings together by force.[38] Necessity and reason are held together not by reason and its appeal to a unifying principle, but by manipulation, and, therefore at best, by a kind of forceful reasoning. There is not one truth to which necessity should finally be docile (for example "the" best or "the" good), nor one principle that will unite. If there is a logos that engages origins, this engagement will not happen as a return to a primordial unity but as a struggle that takes place in the difference between the two beginnings. As John Sallis points out, the meeting point of the two beginnings would involve a certain "hostility, as when two soldiers stand together face to face in battle, in close combat with one another."[39] The two beginnings are held in a tension that enacts the difference out of which and in which events of beings come to presence as sensible and intelligible beings. The beginning of all events of beings occurs out of a difference marked or figured in Timaeus' story by *chora*. This is a figure of a kind that is hardly a kind (in the metaphysical dualistic sense), and that, as we will see in the next sections, will remain in its strangeness a figure of birth and possibility for the occurrences of beings.

Chora: The Nonpresence and Alterity of Events of Beings

The starkest word we encounter concerning Timaeus' second beginning, particularly in introducing *chora*, is "difficult." Timaeus begins and ends the sentence where he speaks of grasping *chora* with the word *chalepon* (difficult).[40] *Chora* is obscure (*amudron*),[41] almost not believable (*mogis piston*), and reachable only by a bastard thought (*logismo tini notho*).[42]

36. *Peithein* comes from the verb *peitho,* "to prevail upon," "to win over," "to persuade" (Homer), also "to mislead" (*Odyssey*) and "to bribe" (Herodotus). In the middle voice it means "to submit oneself" (Homer). In the passive it means "to trust" and "to rely on" (Homer). In Plato's *Sophist* it means "persuasion by fair means," and in its passive form "to be persuaded to do." See Sallis, *Chorology,* 91–98.

37. Klein, *A Commentary on Plato's Meno,* 194–95.

38. Cf. *Timaeus*, 56c.

39. Sallis, *Chorology,* 93.

40. *Timaeus* 49b, "chalepon de allos . . . chalepon." This is what Sallis elegantly calls Timaeus' "double warning" concerning *chora,* in "Timaeus' Discourse on the *Chora.*"

41. *Timaeus*, 49a4.

42. *Timaeus,* 52b2.

The foremost characteristics of *chora* are neither a matter of simple appearances nor graspable in terms of objective presencing. *Chora* is a matter of difficulty. Timaeus' emphasis on difficulty recalls the foreign character, the strangeness of *chora* as a "third kind," and in this way also points to a way to engage the matter. The issue is not graspable with the same certitude of the first account, i.e., by beginning from the supposition of a self-identical cause written on the heavens and reachable through reason, and by subsequently searching for such an ordering principle.[43] Rather, the strangeness of *chora* will call for a second beginning that has only the sole certainty of the difficulty of grasping the events of the arising of beings that the term figures.

When one listens to the difficulties characteristic of *chora*, one finds that the word marks the impossibility of grasping the origins of events of beings in terms of either objective or material presence. The conceptual uniqueness of *chora* as used in the *Timaeus* becomes strikingly clear by contrast with the usage of the term both in Homer's language and in ancient Greek and the ancient world in general. In this tradition the word marks "the space in which something is," a "place."[44] This is the general sense also found in other works by Plato, who in the *Laws* uses it as synonymous with the place of things, humans, and cities, i.e., as terrain, landscape, or country.[45] As well, this sense of the term appears with Aris-

43. *Timaeus*, 37d.

44. Liddell and Scott, A *Greek-English Lexicon.* Cf. "Region," *lieu, Ort, Stelle, Distrikt, Raum,* in P. Chantraine, *Dictionnaire Etymologique de la Langue Grecque* (Paris: Klicksieck, 1980) and Frisk, *Griechisches Etymologisches Wörterbuch* (Heudekberg: Universitätsverlag, 1973).

45. An interesting ambiguity appears in the *Critias,* where Socrates gives an account of the city of Atlantis and then supplements his account by speaking "of mountains and other spaces [*ton horon kai tes alles choras*]" (Cambridge, Mass.: Harvard University Press, 1989), 119a. What is particularly significant is a certain ambiguity that appears at the edge of the city and is marked by *chora,* i.e., by Socrates' inclusion of the indeterminate spaces outside of the city in his account of the city. In contrast to Pierre Leveque and Pierre Vidal-Naquet, I find the line from the *Critias* not "mysterious" but telling in its ambiguity. Already in this passage the *chora* marks a place outside the economy of the dianoetically generated city, and, at the same time, it marks the city in its otherness. See Leveque and Vida-Naquet, "Space and the City: From Hippodamus to Plato," in *Cleisthenes the Athenian,* tr. David Ames Curtis (Atlantic Highlands, N.J.: Humanities Press, 1996), 90.

One path to follow regarding the question of the relationship between the *polis* and *chora* would be to reread the *Republic* in light of the double account of the cosmos in the *Timaeus,* and in light of how this doubling stops the unquestioning dianoetic reading that would understand the *polis* either in terms of production or of the logical causality of objective presence. Of course this is an obvious suggestion in view of the fact that the account of the cosmos in the *Timaeus* occurs as a continuation of the making of a city in speech in the *Republic.*

totle's equation of chora and "place" (*topos*) in Book IV of the *Physics*. This is also the sense that is passed down as *chora* is translated into Latin. Calcidius translates *chora* as *locus*.[46] Simplicius, in his commentary on Aristotle's *Categories*, translates *topos* as *locus*.[47] The result is that the strangeness and difficulty of the question of being as figured by Timaeus' *chora*, as well as the difference between this term and the place (*topos*) of things at hand, are finally lost. Whichever term one chooses, they all refer to one particular kind of "spatiality," namely, to the spatial relation between things present at hand. In terms of presence, spatiality becomes objective space. Things can be next to each other, or inside one another. "Space" may also reach to the limiting edges of a determinate thing, and as such it is the place of things. Furthermore, "space" is also a vessel-like container, an abstract whole wherein all things take place. This sense of space is heard when one speaks of being "in the world." This last sense of space takes us to the ideal or transcendental sense of space. This is a sense of space that must already be in place so that we are able to experience things "in" space, their spatiality of place, and, in general, their taking place. One might say that, as containing all beings and as facilitating the possibility of experiencing these beings' taking place, the vessel has now become ideal. These ways of interpreting spatiality are as common in everyday discourse as they are in scientific, metaphysical, and transcendental discourses. However, when Timaeus introduces *chora* in his second account, this word marks a sense of spatiality outside such conceptions of spatiality in terms of objective and ideal presence.

The difference between *chora* and spatiality interpreted in terms of objective presence (*topos*) appears in Timaeus' own introduction of *chora* as the "receptacle" of all becoming (although the introduction seems at first another objectifying interpretation of the matter). "[B]ut now the argument seems to compel us to try to reveal by words a Form that is difficult and obscure. What essential property, then, are we to conceive it to possess? This in particular, that it should be the receptacle, and as it were the nurse, of all Becoming."[48] *Chora* is called a "receptacle" not in the sense of being a container, a thing that contains, but because of a certain receptivity.[49] According to Timaeus, *chora* is *a-morphon*: without a determinate

46. Sallis, "Timaeus' Discourse on the *Chora*"; and *Chorology,* 115.

47. Simplicius, *Commentaire Sur Les Categories D'Aristote,* tr. Guillaume de Maerbeke (Leiden: Brill, 1975). From *locus* we get the terms "location," "locality," and "local."

48. *Timaeus,* 49a.

49. Klein, *A Commentary On Plato's "Meno,"* 198–99.

form or shape, *chora* is and must be indeterminate.[50] Why? It is precisely this indeterminacy that is the operative element in *chora*'s receptivity, since it is in the absence of a determinate and unchanging form that events of beings will take their unique and diversifying forms. "Wherefore it is right that the substance which is to receive within itself all kinds should be void of all forms. So likewise it is right that the substance which is to be fitted to receive frequently over its whole extent the copies of all things intelligible and eternal should itself, of its own nature, be void of all the forms."[51] Only indeterminacy can allow for all beings to take form, to become determinate in their singular diversity. This means that if *chora* is understood as the receptacle of all beings it must be an event of spatiality outside of presence, an event out of which all determinations of beings occur but that will itself remain outside determination. On the one hand, this means that *chora* appears as an essential aspect of events of beings that remains beyond objective presence. Furthermore, the issue is no more a matter of conceptual representation than it is one of objective presence. That *chora* must remain indeterminate also means that, on the other hand, there will not be any possible representational discourse that will make this matter wholly intelligible. In this sense, *chora* remains beyond the assertive logic of those discourses grounded on ever-present, unchanging first principles or origins, as well as those grounded on objective facts.

Perhaps, in light of this indeterminacy, one can say now that rather than being a term that defines a kind or form, *chora* is a figure that echoes something of presence and is yet beyond presence and representation, something strange, a certain *ektopic* aspect (*ek-topos:* what is other, strange) of the taking place of events of beings.[52] This figure indicates a certain duality in events of beings: all visible or intelligible events, including thought, will come to presence in light of a certain withdrawal or absencing essential to them. This duality of events of beings points to a difficulty central to the philosophical logos. In this relation to *chora* the logos will not only be a matter of presence but also of engaging the withdrawal essential to events of beings. The issue here is not withdrawal as

50. *Timaeus,* 51a. See also Sallis, "Timaeus' Discourse on the *Chora.*"

51. *Timaeus,* 50d–51a.

52. In Sophocles, *oudenos pros ektopou,* "by no strange hand"; and, in Aristophanes, "out of the way," "strange," "extraordinary" (Liddell and Scott, *Greek-English Lexicon*). I point to this word in contrast to *a-topos,* which is used for being out of one's place, and in order to suggest a sense of spatiality that is outside that of a topology of presence and, as such, cannot be reduced to or sought after in terms of presence.

nothing, mystery, or obscurity. The point is that in light of the doubling of events of beings, the logos will be called upon to engage specifically and rigorously all objective and ideal presences in their events, and this means in their presencing and absencing. Here the philosophical logos comes into question: how will thought occur outside a discourse grounded on objective and ideal presence, unchanging origins, and the task of representation, in order that this absencing begins to be engaged? The question takes one back to Timaeus' account. How is one to interpret this account in light of this difficulty? How and to what extent does Timaeus' discourse engage this difficulty? The latter questions are twice invited by Timaeus himself, once when he points to his discourse and puts it in evidence by naming it a "likely story," and a second time, when he interrupts his discourse at the limit of dianoetic possibility and begins again by introducing *chora*.

Speaking of Beings: Presence, Nonpresence, and Alterity

Thus far this discussion of Timaeus' likely story has examined the issue of the strangeness of events of beings as figured by *chora*. In order to begin to engage the difficulty of the alterity of thought and of events of beings, one must now follow Timaeus to the limits of what can be said in terms of sensible-intelligible presence. As Jacob Klein points out, Timaeus' greatest difficulty in telling his likely story is found in his attempt to give us an originary account of the cosmos in the language of what is already generated. "For our *dianoia*, manifested in our speech, is indeed turned toward this familiar world of ours, turned in the main, toward visible bodies, which our *dianoia*, in the exercise of its power of *dianoetic eikasia*, understands as 'copies' of what is intelligible only."[53] The figure of *chora* brings forth a paradox that places Timaeus' likely story between a language of sensible-intelligible presence (*dianoia*) and what cannot be grasped in terms of visible, sensible, or even intelligible presence understood as representation.

Klein's passage indicates that Timaeus' likely story (*logon ton eikota*) is intended as a copy.[54] Literally, the account is an *eikon*, a "copy" in speech, an "image" through which the cosmos is seen. This interpretation

53. Klein, *A Commentary On Plato's Meno,* 199. Cf. F. M. Cornford, *Plato's Cosmology* (London: Routledge, 1952), 197: "He then passes to a description of the Receptacle and its contents, imagined as existing 'before' the ordered world came into being."

54. *Timaeus,* 29b; cf. 48d; see note 18.

of the logos as image points to the engagement of events of beings in terms of presence alone. Both of Timaeus' accounts are sustained by thinking of being as presence, and they are delivered in a language exclusively of presence. Timaeus' story is an image of the origin of all sensible beings. At the same time, this image represents the origin by producing a copy of an ever-present order or being.[55] Here it is both through and as a copy of what is ever-present and unchanging that the sensible is known in its intelligibility. The idea of being as presence that pervades Timaeus' account is evident in its first part, in which he draws a distinction between two kinds or forms, being and becoming. This distinction is grounded on a double idea of being as presence. In the story being and becoming are called *eide*, or forms. *Eidos* means "form" in the sense of visible shape; it is derived from the verb "to see," (*horao*).[56] All of being is understood in terms of visible presence. Being is an intelligible *eidos*, a form that is uncreated, a being continuous and self-identical, and in this self-identity an ever-present determination. Becoming is blindly present in its coming into being and passing away.[57] Indeed, presence is always operative in the differentiation between objective and ideal presence, a constant indicated by the very word for kind or form (*eidos*). This is the case even when one follows traditional rationalist and metaphysical interpretations of events of beings, and says that reason as knowledge of the intelligible outweighs the senses, and analogously, that truth outweighs opinion, and that being outweighs becoming.

The understanding of being as presence extends to the second part of Timaeus' account. In spite of the characteristics of *chora* as an event outside the dianoetic logos or ordering, and in spite of its formlessness, Timaeus refers to *chora* as a form or kind (*eidos*). This emphasis on objective and ideal presence and its determinacy in terms of form or idea is also evident when Timaeus says that it is because the four elements are not always a self-identical presence and keep changing that they must not be addressed as "beings" but rather as "such-like": Only that which is unchanging in form and ever-present, hence determinate and determinable, can be called "being," a "this" or a "that."[58]

55. *Timaeus,* 37d.

56. As A. E. Taylor points out, "[t]he logos or discourse of the soul is sometimes concerned with what is perceptible to sense, sometimes with what is not; in both cases it is possible to have a true discourse." Taylor, *A Commentary on Plato's Timaeus* (Oxford: Clarendon Press, 1972), 178.

57. This is illustrated in Socrates' story of his shift from the blinding study of the *phenomena* toward the logos in the *Phaedo* (Cambridge: Cambridge University Press, 1993), 96c.

58. *Timaeus* 48b–50a; *Timaeus,* 29a.

Timaeus does not abandon the language of presence in his inquiry, his likely story. In his account he separates language from *dianoia*.[59] It is only when *dianoia* puts language to its service that something of the Being of beings is said. *Dianoia* means "through *nous*," and *nous* is the direct apperception of the ever-present forms, or being. Language functions as a mimetic tool that sounds out the intuited forms given in and experienced through *nous*.[29,60] This is what is already indicated in the start of his account when he presents it as a likely story, a copy in speech of what is, and this means that language is understood as having meaning in what is being copied. The logos is a kind of snake that wraps tightly around the body of self-identical truth, and generates meaning by clenching to this ever-present, unchanging center. Timaeus' thinking relentlessly holds on to presence for its meaning, letting go of becoming but not of the ever-present forms. This thinking of presence engages language as image-making and as making present what cannot be grasped by opin ion or the senses.

However, as we have seen, *chora* is beyond representation. Characteristically, *chora* must remain always indeterminate, and this lack of determination places it altogether outside of the language of determinate presence, either in terms of an ever-present being or in terms of the formal and material processes of the becoming of things. The characteristic indeterminacy of *chora* also suggests that one cannot hold on to the metaphysics of presence in order to think the origin of beings. Because *chora* is outside of presence it does not make sense to interpret it in terms of presence. In terms of presence *chora* is meaningless, nothing, a kind of no place (*a-topos*).[61]

59. Klein, *A Commentary On Plato's Meno,* 198–99.

60. One should keep in mind that there is a difference between the modern understanding of intuitions (i.e., Descartes) in the mind, and the way *nous* is given in concrete experience in Greek thought.

61. Once again with this Greek term I am suggesting a differentiation between *atopos,* the Greek word for a place that is not at all, a "space" that draws its nonsignificance from the characterization of space in terms of thinghood; and the possibility of understanding place outside the interpretation of metaphysics, as *ektopos,* the Greek word for what is "strange."

A second point to keep in mind is that *a-topos* suggests a certain disengagement from the density and activity of beings in their *phusis,* and as such *a-topos* runs the risk of once again transcendentalizing events of beings—events that, as the word *phusis* indicates, occur in the coming to be in passing away of their essential temporality. See Heidegger, "Aletheia (Heraklit, Fragment 16)," in *Vorträge und Aufsätze* (Pfullingen: Neske, 1985); and Heidegger, *Early Greek Thinking,* tr. David Farrell Krell and Frank A. Capuzzi (New York: Harper, 1975).

But *chora* is not nothing: *Chora* is essential to events of beings. At the same time *chora* figures a rupture of the dianoetic interpretation of being. When Timaeus introduces this figure he interrupts the dianoetic interpretation of beings, the dianoetic logos. In doing this, he puts any interpretation of the origin of the cosmos beyond an ever-present and self-identical being, and interrupts the function of language as the mimetic tool of such ever-ruling presence. The point here is not that language is imprecise or insufficient with regard to truth or ever-present forms. *Chora* figures an essential element of the delimitation of truth and all events of beings in logos. Traditionally, Timaeus' second beginning is interpreted according to unchanging principles and their mimetic logos: this occurs in our struggle to find a unified origin of the cosmos, and to do so according to the language of presence, in our attempts to give a positive, or visible, account of *chora*. Here one finds Timaeus' discourse at the limit of the senses of being: if not in terms of objective and ideal presence, if not as sensible or intelligible, how will one begin to understand the figure of *chora*? When one encounters this interruption—the impossibility of posing the question of being according to objective and ideal presence—philosophical discourse must come to a stop. There is no path to follow, no structure, no image; one lacks the "truth" by which discourse traditionally secures meanings. And yet it is this interruption that echoes in the speaking and the sense of events of beings articulated in Timaeus' second account.

Chora: From Signification to Intimations of Alterity

With the interruption of the dianoetic interpretation of the cosmos comes a certain unsettling of an order ruled by ever-present, unchanging first principles and everlasting, unchanging forms. If the "meaning" of beings—or, in other words, being as such—is understood exclusively in terms of the metaphysics of presence, one might conclude that with Timaeus' introduction of *chora*, of a kind indeterminate and beyond kind, the end of "meaning" is announced. In the loss of unchanging origins the events of beings are left adrift, bereft of meaning, direction, form. However, what I have shown thus far is not that *chora* indicates such a determinate end or radical nihilism. The loss of the centrality of first principles guided by an overarching, unchanging rule only indicates a certain possibility for meaning that exceeds or goes beyond the limits of Timaeus' first account or dianoetic cosmology (i.e., presence and representation). What has been lost is not the sense of beings but the interpretation of events of beings in terms of meanings determined by ever-present metaphysical principles. Here a certain transformation of the sense of the philosophical logos occurs, from likely stories to another way of engaging events of beings. The strangeness of events of beings as figured by *chora* calls for a thinking that understands being other than in terms of a logic of presence. In the collapse of a purely dianoetic account of beings the question of the sense of beings is separated from that of signification. Up to a certain point in Timaeus' account, beings can be the matter of likely stories, as we have already seen, a term that posits the interpretation of the philosophical logos as an account or representation of beings, a re-presentation of their unchanging "origins" and of their first unchanging principles and order. Likewise, the logos is also interpreted as a kind of image-making of this unchanging order, a mimetic device. However, now, in the impossibility of sustaining this interpretation of the occurrence of beings in terms of objective and ideal presence, the very task of the philosophical logos comes into question. In other words, the collapse of the dianoetic order that occurs with Timaeus' introduction of the figure of *chora* intimates the question of being by breaking away from first principles and through the release of the reductive understanding of language as story-telling or image-making. Here the difficulty and interruption of Timaeus' second beginning intimates the need for another way of thought.

The figure of *chora* brings the logos to its limits, and at the limit the question of the senses of beings beyond objective and ideal presence is discovered. In light of the collapse of the interpretations of the senses of beings in terms of presence, this presupposed ground and its representational logos must be suspended: all that remains for thought is the interruption . . . one must stop, and then begin again. But, now, one begins from another beginning (echoing Timaeus' second beginning), a beginning that recalls the question of being to the difficulty of a sense of beings beyond objective and ideal presence, and beyond the sensible-intelligible world as figured by such a metaphysical dualism.[62] This second beginning

62. This aporetic moment in Timaeus' account echoes two other moments, two other beginnings of the philosophical logos. One is the interruption marked by the passage from the *Sophist* at the beginning of *Being and Time.* The other is Socrates' account of his beginnings as a "second sailing" in the *Phaedo* (95a–102a). On both occasions the logos takes its course in light of the alterity of the discourse as marked by its interruption and by a leap to another way of thought. In the case of the second sailing, it is in light of blindness that Socrates finds his way.

occurs from alterity, out of senses of beings that are not grounded on presence alone and that are therefore not representable, hence that are other, strange to the interpretations of being as presence alone. Thus, Timaeus will have to refer his listeners to *chora* as a "third kind," beyond the duality of objective and ideal metaphysics. This is a beginning at the limit of the sense of being, since it begins in the collapse of the interpretation of events of beings in terms of objective and ideal presence, and because it intimates a certain need for another way of engagement of events of beings. Here the logos moves beyond the edge of the sensible and intelligible world, and finds itself adrift, open to delimitations of beings well beyond the conceptual determinations of metaphysical dualism, i.e., the two dianoetic kinds, and here one encounter a beginning without a presupposed path to follow (without a fixed teleology), one found neither in idea nor in fact. At the same time Timaeus' second beginning calls for thinking, a logos, that cannot be interpreted by recalling it to presence. It will not be a matter of affirming the mimetic function of signification that gives meaning to the dianoetic logos. Rather, the need for thinking—a task that Timaeus will relentlessly remain with—indicates a path outside presence, an "errant path," a "bastard thinking," as Timaeus himself says.

Chora: Enactments of the Alterity of Being

In the loss of a grounding of thought in presence, the task of the philosophical logos has changed from a question of mimetic signification to that of the senses of events of beings. But how does such a shift occur in Timaeus' discourse, a logos that is by definition a likely story, i.e., a discourse meant as a mimetic event of signification grounded in ideal presence? This question can be posed in a more threatening form both for this discussion (putting in danger its claim that there is such an interruption to the mimetic function of the logos) and for the interpretation of the logos as a mimetic event (by opening such an interpretation to aspects of Timaeus' discourse that interrupt and violate this mimetic account of Timaeus' logos). Does Timaeus' account indicate more than what it intends to say? How does this occur?

As the discussion indicates up to this point, the essential alterity of the occurrences of beings as events beyond presence and representation is felt in the difficulty and indeterminacy of these events figured by *chora*. We are speaking of the impossibility of grasping and giving determination to this figure through a representational account. This difficulty indicates a certain withdrawal beyond presence essential to beings in their coming to presence, a withdrawal (*choreo*) in light of which events of beings arise, and that will remain outside presence and beyond representation. How is this aspect of events of beings engaged in Timaeus' likely story?

The interruption of the smooth continuity of a logos of presence and representation is not announced as such, not presented, not discussed at all. It is by way of an interruption that remains beyond Timaeus' representational discourse that the figure *chora* is introduced (the break between the two accounts and the difficulties we have been discussing are never a thematic focus of study for Timaeus). This interruption is covered over by Timaeus' own emphasis on presence, although a closer look at this likely story, in light of the interruption, will offer a sense of the essential alterity of the events of beings as intimated by the introduction of *chora*.

Timaeus will persist in his attempt to give an account of the occurrences of beings even after he has come to the limit of the dianoetic ordering of all beings.[63] However, *chora* is not operative as an image of something. As such it may only indicate a certain almost imperceptible indeterminacy, an odd presence, never quite graspable,[64] a being almost unbelievable, and an errant motion, which together call for a bastard thinking. As these same terms indicate, when taken up without emphasis on presence and signification, *chora* composes an experience of irreparable difference, i.e., all senses of being are recalled in their essential withdrawing in coming to presence, a withdrawing that, in marking the being of beings beyond presence and representation, recalls thinking to the essential alterity of the events of beings, as well as to the alterity of the philosophical logos.[65] In spite of Timaeus' emphasis on presence, his intended likely story goes well beyond representation. The radical aspect of his discourse comes powerfully to light when one considers the difficulties brought forth by

63. Much can be made of the difference between the two accounts in terms of a difference between reason and poetry. However, this difference only indicates the inseparability of the two that is operative throughout Plato's dialogues. In light of this inseparability, one cannot explain away the difficulty figured by *chora* by appealing to reason and its distinct "other."

64. *Timaeus,* 51a.

65. All of this is not a discussion of something, not even an image; it is what *chora* suggests, an opening of coming into being, indeterminately going on and relinquishing its position in the most powerful way for the sake of determination. It is a giving space that is not at all the spatiality of any determination, of any particular presence.

his second beginning regarding the primary role of kinds or forms and of reason as foundations for the discourse.

Timaeus' claim that *chora* is a "form" or "kind" (*eidos*)[66] has a deformative effect for the traditional dianoetic image or interpretation of the cosmos.[67] According to this claim, first of all the very concept of "kind" will have to be engaged not only in terms of presence but also in light of the withdrawing and indeterminate openness figured by *chora*. Second, thinking will have to turn to an engagement of beings beyond the metaphysical or dianoetic interpretation of beings in terms of objective and ideal presence, as well as beyond the interpretation of the world as an intelligible and sensible presence. Third, in view of the irreducible difference between self-identical cause and "errant cause," the fact that Timaeus calls the *chora* a "kind" suggests another truth than that of reason oriented by the logic and teleology of objective and ideal presence. This is also clear in the way that the two causes essential to the events of beings are held together according to Timaeus' account.

According to Timaeus, reason persuades necessity by force. It is not an ever-present, unchanging logos that leads the two causes into one.[68] Ultimately Timaeus' account cannot appeal to such first principles. Truth, as the gathering of the two causes in events of beings, will always be a matter of reason being called beyond itself by having to force necessity, an enactment of force that will always find reason as the follower of necessity in its errant course. This slipping of truth figured in the passage of all beings is also figured by the way all beings come into being according to Timaeus' story, and significantly, in his dianoetic account. Once again, it will be by force that beings come to be. The cosmos is generated in the dianoetic account out of incompatible elements mixed together not by agreement, but by compulsion, by force (*bia*): "by forcing the other into union with the same, in spite of it being difficult to mix."[69] In short, if *chora* is a kind beyond kind, then has truth not gone beyond itself as presence unified by reason? Does one not find

66. Timaeus calls *chora* a "third form." But as such, the figure is a third kind wholly other than being and becoming. This "otherness" is itself difficult. Because *chora* is outside the order of the dianoetic logos, it cannot be thought in terms of its categorization. Therefore *chora* cannot be thought to be in the relation of the first two kinds to each other. The otherness of *chora* means utter difference from being and becoming.

67. Sallis, "Timaeus' Discourse on the *Chora.*"

68. See notes 20–21.

69. *Timaeus*, 35a. *Bia* means "force," "power," or "might." The verb *biazo* means "to be overpowered," "to be overcome by violent force."

in Timaeus' likely story intimations and echoes of truth as an errant course beyond presence, and a call for a bastard thinking beyond representation?

Chora: Conclusion

Timaeus' story leads us beyond the sensible-intelligible world of objective and ideal presence. In terms of spatiality one can say that *chora* marks the opening of the issue of spatiality outside objective and ideal space. However, this figure goes much further. Timaeus' likely story leads us to the limits of objective and ideal interpretations of the senses of beings, and in doing so intimates a certain opening toward thinking events of beings and thought in their alterity. This occurs as *chora* remains both essential to events of beings and beyond representation and presence. At the same time, this withdrawal indicates a limit of representational mimetic discourse, and in this way points to the need for engaging events of beings through a wider philosophical logos, one that will engage the withdrawal essential to them.

Aristotle's *Topos*

In Aristotle's *Physics* we find a linking passage between Plato's *Timaeus* and Heidegger's project in *Being and Time* (the reawakening of the question of being). In Book IV of the *Physics* Aristotle takes up the figure of *chora* in the *Timaeus* as a point of departure for his discussion of spatiality as place (*topos*). In *Being and Time*, Heidegger begins his analysis of the question of being (the dasein analytic) by paraphrasing Aristotle's discussion of *topos* in the *Physics*. Aristotle's discussion is guided by objective and ideal presence, and he interprets Timaeus' difficult figure in terms of the logic of presence. Unlike Timaeus' likely story, Aristotle's logos aims to give a precise account of *chora* by interpreting it in terms of objective-empirical observation and through a *logos apophantikos*, which will make evident the occurrence of beings in terms of categories, causes, and first principles. This assertive logic of presence is meant to take the place of Timaeus' account and to overcome the shortcomings of Timaeus' story-telling. The question is: what happens to the essential difficulty figured by *chora* when taken up by Aristotle's logos? How will the nonpresence and the impossibility of giving a representation of *chora* be engaged or received by Aristotle?

Beginning From Chaos, Before Timaeus' *Chora*

Aristotle's term for spatiality is "place" (*topos*). The term appears in Book IV of the *Physics* and indicates a rethinking of Plato's *chora*. According to Aristotle, among his predecessors only Plato in the *Timaeus* addresses the issue of spatiality: "[W]here everyone asserts that place is, only Plato has attempted to say *what it is* [*ti d'estin*]."[70] Therefore, Aristotle's discussion of spatiality begins by following Plato's lead and by making the question of the "whatness" of place (*topos*) the leading question of his inquiry. "The Natural Philosopher has to ask the same questions about 'place' as about the 'unlimited'; namely, if it exists or not, in what way is it, and what is it [*ei estin me, kai estin*]?"[71] As we will see, this translation of *chora* into place covers over the difficulties figured by Timaeus' term, while it also points to the limits of the logos of presence as found in Aristotle's *logos apophantikos*.

In order to give direction to his inquiry into the question of place (*topos*) Aristotle turns to a cosmology prior to that of Timaeus, to another story concerning the beginning of events of beings, to Hesiod's *Theogony*:[72] "[T]his would justify Hesiod in giving primacy to Chaos where he says: 'First of all things was the opening [*chaos*], and next broad-bosomed Earth'; since before there could be anything else there must be room [*chora*] to be occupied. For he accepted the general opinion that everything must be somewhere and must have a place."[73] Here, Aristotle equates the questions of *chora* and *topos* with primordial *chaos*,[74] the gaping void out of which all beings come into being in this myth of creation.[75] Both this myth and Timaeus' story orient the ques-

70. *Physics,* IV: 1, 209b17.

71. *Physics,* IV: 1, 208a27–29. Cf. 209a4–5; 209a30.

72. Hesiod, "Theogony," in *The Homeric Hymns and Homerica,* tr. H. Evelyn-White (Cambridge: Harvard University Press, 1936), 78–153.

73. *Physics,* IV: 1, 208b30. The quote is from Hesiod, "Theogony," line 116; cf. line 699.

74. Here origin, or *genos,* appears not as an ever-present being but as *primus-ordiri,* first created.

75. The first instance of *chaos* means a gaping void, an empty but created space. "'Chaos,' von Hes. Th. 116 als Bez. des Erstentstandenen gebraucht, gewöhnlich (seit Arist.) als ein leerer Raum aufgefasst." H. Frisk, *Griechisches Etymologisches Wörterbuch,* vol. 2 (Heidelberg: Universitätsverlag, 1973). As Sallis points out, the story of *chaos* told by Hesiod is also figured in the *Timaeus. Chorology,* 85.

tion of place for Aristotle, and yet, in doubling back to Hesiod, Aristotle moves away from the likely story in the *Timaeus* and toward another kind of account, a logos that will attempt to overcome the indeterminacy of *chora*, and that in the attempt will cover over the difficulty found in the figure of *chora*.

When one looks at the first lines of Hesiod's poem as quoted by Aristotle, one finds that Aristotle's association of *chora*, place (*topos*), and *chaos* figures a change in the approach to the difficulty introduced by Timaeus' likely story. In the poem, *chaos* is created. Hesiod uses the term *genos* for being and not *hen*: thus *chaos* is not an ever-present being but what is created first, and therefore it is "primordial" in its literal sense, *primus-ordiri*, first to begin. This means that *chora* and *chaos* have in common the character of an originary opening.[76] As such, *chaos* has the character of being a gap between earth and sky that is neither transcendental nor ever-present. However, at the same time, the fact that *chaos* is a created being suggests that this opening is a kind of being, a something that can reply to the question "what is it?" (*ti estin*).[77] In short, by equating *chora* with *chaos* Aristotle suggests that, although these figures indicate a primordial opening necessary for beings to be, this opening is a kind of place, i.e., a particular being that can be studied and given determination. This is in fact precisely what Aristotle praises about the *Timaeus*, that in Timaeus' account *chora* is engaged as a kind of being. This particular shift in interpretation, from *chora* as a difficulty beyond presence or representation to this figure as entity, marks a path that moves away from the withdrawing aspect of *chora* toward the interpretation of the spatiality of beings in terms of presence alone. Aristotle directs his inquiry by asking whether place exists, in what way it is, and what it is.[78] Furthermore, in Book IV, Aristotle states that the first step in his inquiry is to determine the

76. "Chaos aus 'chaos' verhält sich zu 'chau-nos' wie das in dieselbe Begriffssphäre gehörende 'erebos.'. . . Da für 'chaunos' eine Grunbedeutung. 'locker, löcherig, mit Löchern versehen' am nächsten liegt, würde sich für chaos eine (relativ) ursprüngliche Bedeutung wie etwa 'Loch, Hohlraum, leerer Raum, klaffende Öffnung' ergeben. . . . Weiteres s. 'chasko' und 'chora.'" Frisk, *Griechisches Etymologisches Wörterbuch*.

77. Casey, *The Fate of Place*, 7–11.

78. This set of questions delineates Aristotle's systematic approach as introduced in his *Organon*, where the questions of knowledge are: what a name means, whether or not a thing corresponding to the name exists, what it is, what are its properties, and why it has these properties. See David Ross, *Aristotle* (New York: Routledge, 1995). See also Aristotle, *Posterior Analytics* (Cambridge, Mass.: Harvard University Press, 1987), II: 1; and II: 71, 9b–72a7.

genos,[79] the kind of being of place. Although Aristotle's discussion sets out from Timaeus' likely story and Hesiod's account, these beginnings are replaced by another kind of logos, one that reaches back toward the *muthos* (poetry) of Hesiod's poem only to bring the indeterminacy of *chora* into the objective and logical light of presence by giving determination in speech to this difficult figure.

This is a bold proposition in light of the difficulty (*chalepos*) that Timaeus encounters in attempting to give a likely story, in attempting to bring *chora* to presence in his account. As we saw in the *Timaeus*, *chora* figures spatiality beyond and outside the dianoetic order of the cosmos, and as such it is "a kind of kind beyond kind, a kind of kind outside of kind."[80] *Chora* is an event of nonpresence, hence it escapes the language of presence. This brings us to the question, to what extent will Aristotle's logos engage Timaeus' *chora*?

Topologies

In the *Physics* Aristotle deals with the existence of things in a way altogether different from that of the *Timaeus.* Aristotle attempts neither to generate the cosmos out of forms nor to tell a likely story. Rather, he takes the inverted path, attempting to arrive at first principles by looking at the phenomena.

For Aristotle the question of the existence of beings is not concerned with forms (*eide*) outside becoming, but with *energeia,* the "being-at-work" that occurs in the activity of life.[81] In short, being is ever-present activity. This drive for the activity of life leads Aristotle to begin with the phenomena. In the *Metaphysics* he asserts that in order to know the existence of beings one must know their primary causes, and these are brought to light not in the *Metaphysics* but in the *Physics.* "It is clear that we must obtain knowledge of primary causes [*arches aition*], because it is when we think that we understand its primary cause that we claim to know each particular thing. Now, there are four recognized kinds of causes. . . . We have investigated these already in the *Physics.*"[82] In a sense,

79. "We must begin by determining to what category [*genos*] it [place] belongs." *Physics,* IV: 1, 209a4.

80. Sallis, "Timaeus' Discourse on the *Chora.*"

81. Jacob Klein, "Aristotle, An Introduction," in *The Lectures and Essays of Jacob Klein* (Annapolis: St. John's College Press, 1985). Klein cites the *Metaphysics,* Theta 8, 1050a21–23; and *On the Soul,* Beta 4, 416b3.

82. *Metaphysics,* I: 2, 983b.

the ground for Aristotle's *Metaphysics* is his *Physics,* where principles and causes are revealed in their activity.

Aristotle's inquiry concerning place follows such a path. At the beginning of the *Physics* he explains that in order to arrive at first principles one must begin with the phenomena, with what is most familiar, and only then move to what is intelligible in its own principles. Elements and principles are only knowable through the analysis of what is accessible to the senses (*aisthesis*). "Now the path of investigation must lead from what is more immediately cognizable and clear to us, to what is clearer and more intimately cognizable in its own nature. . . . [T]he things most obvious and immediately cognizable by us are concrete and particular, rather than abstract and general; whereas elements and principles are only accessible to us afterwards, as derived from the concrete data when we have analyzed them."[83] Accordingly, Aristotle articulates all spatiality by referring to the phenomena. He begins Book IV by pointing out that all things that are said to be are said to be "somewhere." Consequently, the question for Aristotle is: What is this "somewhere," this place (*topos*) of entities at hand?[84]

Although he begins his discussion of place solely with regard to the presence of the phenomena, place is for Aristotle altogether different from an entity that occurs in this space: "[T]herefore its 'place' cannot be either a factor or an intrinsic possession of a thing, but something separate in its being."[85] Place is also different from each of the four causes of any thing, it is different from any determination of being: "Again, how is one to suppose that place affects or determines things in any way? For it cannot be brought under any of the four causal or essential determinants."[86] Place appears out of a particular activity, out of the being of the phenomena. What reveals place is the motion of entities at hand: "To begin with we should recognize that no speculation as to place would ever arise had there been no such thing as movement, or change of place."[87] Place is also made present by the motion of the four elements: "Moreover the trends of the physical elements, fire, earth, and the rest show not only that locality or place are something, but that they are active [*all'hoti kai echei tina dunamin*]."[88]

83. *Physics,* I: 1, 184a14–25.
84. *Physics,* I: 1, 208a27–33.
85. *Physics,* IV: 2, 209a28; 209b21–23, 209b32–34; and IV: 3, 210b30–31.
86. *Physics,* IV: 1, 209a20.
87. *Physics,* IV: 4, 211a12–15.
88. *Physics,* IV: 1, 208b9–10.

Aristotle's understanding of spatiality purely in terms of presence is apparent in the way the spatiality of beings is revealed for him out of the presence and behavior of the elements, and out of the motion of bodies at hand. One can contrast his understanding of spatiality with that of the *Timaeus,* where not only entities at hand but even the elements that constitute them must be generated and find their possibility in light of a certain indeterminacy that will remain beyond objective and ideal presence and representation. Aristotle's discussion of spatiality remains always within the evidence of presence, always affirming presence by beginning with and returning to it.[89] Aristotle begins from what is at hand in order to make present the first principles that show themselves to our senses (*aisthesis*) in the presence of beings. The *topos* is always understood in terms of presence, and, in this manner and because of Aristotle's association of the two, the difficulty figured by *chora* is always already inscribed within a topology of presence.

Logos Apophantikos

What sustains this phenomenology of presence is Aristotle's metaphysical aim. Although his method begins with the phenomena, his aim is ultimately a set of ever-living principles and causes—literally, that which is not-dying and ever-visible, (*athanatos, aidios*).[90] In the *Metaphysics* he contends that wisdom (*sophia*) is knowledge of certain principles and causes.[91] In the *Physics* he follows this idea of wisdom by deriving the principles and causes of the occurrences of beings out of the activity of life: "Obviously then, in the study of being [*peri phuseos epistemes*] too, our first object must be to establish principles [*ta peri tas archas*]."[92] The aim of the *Physics* is metaphysical, i.e., not so much the examination of the phenomena but the articulation of the ever-living first principles and categories that constitute all that is said to be in its intelligible activity.

89. Another way to put this is that for Aristotle being occurs out of *phusis.* Thus, in the *Timaeus,* in spite of the performative interruption of *chora,* the ontological tone is sustained by the metaphysical understanding of being in terms of forms (*eide*). This is not to say that Aristotle's thought is not metaphysical, but it does show a contrast between the two of them, both in the way that the phenomena of spatiality are discussed and in the way their metaphysical lineages occur.

90. Klein, *Lectures and Essays of Jacob Klein,* 190–91. Klein cites the *Metaphysics,* Lambda 7, 1072b26–1073a13; and *On the Soul,* Gamma 5, 430a18, 23.

91. *Metaphysics,* I: 1, 932a2–3.

92. *Physics,* I: 1, 184a15.

For Aristotle the way to make being-at-work present is through language, or *logos*.[93] Thus, he defines a human being as the being who has *logos*, rather than giving equal importance to *aisthesis* (also a way through which universals are given and knowledge is possible).[94] Beings become intelligible in language, which means that the discovery and analysis of the right word is the uncovering of the meaning of things. But there are many kinds of speech. For example, in the *Timaeus*, as in the taut fabric of all Platonic dialogues, one finds a mixture of various discourses, among them most dramatically juxtaposed a mixture of mythical and mathematical accounts. Furthermore, speeches can be wrong, mere opinion, or false. In short, as Aristotle himself puts it, being is spoken of in many ways. However, for Aristotle there is only one sure path to clarity, and this is the *logos apophantikos*. The *logos apophantikos* brings to presence the hidden principles of things in their activity; it sounds out the structure of what is given to our senses and then only experienced in an indeterminate or opaquely convoluted fashion.[95] One can already hear the character of this logos in its etymology. *Apophantikos* comes from *phao*, or "to give light," and at the same time it bespeaks a "saying": *apo-phemi* means "to say." One might translate the term then as a "de-claring," a bringing into the light by saying.

This apophantic project marks a change in the way thought engages all senses of being. In a sense Aristotle's task is to make intelligible what is ever-present and unchanging behind the all-changing appearances of things. At the beginning of the *Physics* Aristotle notes that when one begins with the phenomena one begins with a "whole," in the sense of an undifferentiated manifold of elements and principles: "[F]or it is the

93. Klein, *Lectures and Essays of Jacob Klein*, 175–80.

94. ". . . [Z]*oon logon echon*." Aristotle, *Politics*, I: 2. On *aisthesis* as a part of truth concerning primary substances see the *Metaphysics*, I: 1, 980a21–981a20; see also the *Nicomachean Ethics*, VI: 1, 4 – 2, 2; and *Posterior Analytics*, II: 19. In these texts at least two ways appear in which sense perception is seen as fundamental to philosophical knowledge: At the beginning of the *Metaphysics*, where the structure of knowledge is such that sense perception stands at the root of the possibility of both art and theoretical sciences; and in the *Nicomachean Ethics*, where *aisthesis* is one of the ways events of being are given to be thought under truth. The structure of knowledge found in these texts as well as in the *Organon* does not present a contradiction to Aristotle's reference to *nous* as the possibility of knowledge, since in all cases we are speaking of the way being can be thought. See Heidegger, "Phänomenologichen Interpretationen zu Aristoteles: Anzeigen der Hermeneutischen Situation," in *Dilthey-Jahrbuch für Philosophie und Geschichte der Geisteswissenschaften*, Book 6 (Göttingen: Vandenhoeck & Ruprecht, 1989).

95. *Metaphysics*, IV: 2, 1003e33. *On Interpretation*, 5, 17a8; 4, 17a2; 6, 17a25. (Cambridge, Mass.: Harvard University Press, 1969).

concrete whole that is the more cognizable by the senses. And by calling the concrete a 'whole' I mean that it embraces in a single complex a diversity of constituent elements, factors and properties."[96] The word "whole" is also used to refer to single elements.[97] In both cases the term indicates what is given to thought by intuition, what calls for thinking and is to be brought to light through analysis. Timaeus acknowledges a similar problem when he stops his likely story in order to begin again, this time not taking for granted but giving an account of the elements. Indeed, Timaeus' problem is even closer to Aristotle. Timaeus follows his observation concerning the need for an account of the elements with a discussion of the misleading use of nouns when we call the elements "beings."[98] This is a faulty denomination because they are changeable, and only the unchanging forms should be called beings. Timaeus' discomfort with the unclear presupposition of the being of the elements leads him to his introduction of the figure of *chora,* his consideration of the indeterminacy that seems necessary for any determination. When this is contrasted with Aristotle, one begins to see the force and direction behind the latter's thought. In Aristotle one finds an intensification of the concern with language. For him the *logos apophantikos* should overcome the indeterminacy of Timaeus' difficulty (*chalepos*). Aristotle deals with this undifferentiated manifold (the elements, the phenomena) given through experience by appealing to language, by trying to clarify the terms one uses in speaking of being. In the *Physics,* Aristotle says, "The relation of names to definitions will throw light on this point; for the name gives an analyzed indication of the thing [e.g. a circle] but the definition analyses out some characteristic property or properties. A variant of the same thing may be noted in children, who begin by calling every man father and every woman mother, till they learn to separate out the special relation to which the terms properly apply."[99] In the face of experience one remains blind to the being of beings, to the activity (*energeia*) or being-at-work (*en-ergon*) that one can only make intelligible as one comes to differentiate principles and causes of beings through an analysis of how they are said to be. In order to arrive at this clarity, in order to give determination

96. *Physics,* I: 1, 184a25–184b.

97. For Aristotle, water and air are considered single wholes, for which moist and warm are attributes. See *Coming To Be and Passing Away* (Cambridge, Mass.: Harvard University Press, 1987), II: 4, 331a20

98. *Timaeus* 48b–50a.

99. *Physics,* I: 1, 184b10–14.

to the manifold, one requires the discipline of Aristotle's logos.[100] (This is not to say that Aristotle is only concerned with language or reason, as, for example, is the case in contemporary analytical philosophy and philosophy of language. It is crucial to keep in mind that logos is a term that does not separate beings from language, but rather that indicates the inseparability of all beings from language in their events or manifestations.) One might say that it will be through a kind of analytical topology that Aristotle's discussion of place in the *Physics* will attempt to give determination to *chora,* beginning with stating what kind of *genos* it is.[101]

In Chapter 3 of Book IV Aristotle discusses place in its active character. The question concerns the many ways things are said to be "in" something (*meta de tauta lepteon posachos allo en allo legeta.*)[102] This does not only attempt to speak of the ways of "being-in" (the relational activity of entities at hand in their being-in in their particular spatiality), it also does so by looking at how this activity of being-in is "said to be." It is the articulation of spatiality by words that concerns Aristotle. The determination of the activity of beings and their spatiality occurs by arriving at logical determination through the clarification of language. This means not only a kind of analytical process but also an analysis of the previous accounts of philosophers, as well as of general opinion. This is the case in Book IV where Aristotle begins by considering the general assumption that all beings are somewhere,[103] as well as in the *Metaphysic* where he speaks of his exhaustive account of causes in the *Physics.* In the latter, Aristotle makes his claim on the ground of the previous philosophers' accounts, a practice characteristic of almost all of his treatises and lectures. "We have given only a concise summary account of those thinkers who have expressed views about the causes and truth, and of their doctrines. Nevertheless we have learned this much from them: that not one of those who discuss principles or causes has mentioned any other type than those which we have distinguished in the *Physics.* Clearly it is after these types that they are groping, however uncertainly. Some speak of the first principle as material."[104] Aristotle concludes by contrasting the uncertainty of previous accounts with the

100. *Historisches Wörterbuch der Philosophie,* vol. 8, ed. Joachim Ritter and Karlfried Gründer (Darmstadt: Wissenschaftliche Buchgesellschaft, 1992).

101. "We must begin by determining to what category [*genos*] it [place] belongs." *Physics,* IV: 1, 209a4.

102. *Physics,* IV: 3, 210a15.

103. "Onta pantes hupolambanousi einai pou." *Physics,* IV: 1, 208a29–30.

104. *Metaphysics,* I: 7, 988a19–25.

clarity accomplished by his *logos apophantikos* in the determination of the four causes in the *Physics.*

Aristotle's Critique of Timaeus' Likely Story

In Book IV of the *Physics,* Aristotle interprets Timaeus' conception of *chora* in terms of his metaphysics of presence, and critiques Plato for his lack of clarity in language. Speaking of the *Timaeus* Aristotle says,

> But if we think of a thing's place as its dimensionality or room-occupancy (to be distinguished from the thing itself as a concrete quantum) we must then regard it as matter rather than as form, for matter is the factor that is bounded and determined by the form, as a surface, or other limit, molds and determines; for it is just that which is in itself undetermined, but capable of being determined, that we mean by matter. . . . This is why Plato, in the *Timaeus*, identifies matter and room, because room and the receptive-of-determination are the same thing. . . . It is no wonder that, when thus regarded—either as matter or as form, I mean—place should seem hard to grasp, especially as matter and form themselves stand at the very apex of speculative thought, and cannot well, either of them, be cognized as existing apart from the other.
>
> But in truth it is easy to see that its place cannot possibly be either the matter or the form of a thing; for neither of these is separable from the thing itself, as its place undoubtedly is.[105]

Aristotle's central criticism is that Timaeus mixes material cause and place, and that as a result of this mistake Timaeus must think of place in terms of the necessary indeterminacy that allows for the formal and material diversity of events of beings.

It is immediately evident from the passage quoted that Aristotle's interpretation of *chora* in the *Timaeus* is grounded on his emphasis on the presence of the phenomena, i.e., the supposition that Timaeus' discussion must be ultimately aimed at the question of place, and ultimately the issue of place must be a difficulty that will answer to the *ti estin* question, a question of whatness or thinghood. It is on this basis that Aristotle interprets Timaeus' account. According to him it is the misunderstanding

105. *Physics,* IV: 2, 209b25.

of the question of place that leads to the difficulty of grasping *chora* and to Timaeus' unclear account. Timaeus' account misunderstands place when it mixes place (*topos*) with the formal and material causes of all beings. This misinterpretation of the problem of place is what must be clarified in order to overcome the difficulty figured by *chora.* The misinterpretation is understandable when one looks at the phenomena. According to Aristotle, in Timaeus' account "room" (*chora*) is "receptive matter" (*metaleptikon*),[106] and this is certainly the way place seems when one looks at the way things seem to "take place." At the same time, Aristotle indicates that when one looks at the phenomena one can easily see that place cannot be either a material or formal cause, since things cannot be separated from their causes as they can from their places.

Aristotle's criticism also is directed to a certain weakness or lack of systematic analysis in Plato's thought. In the *Metaphysics* Aristotle remarks on the limited character of Plato's logic, making direct reference to the account in the *Timaeus.*[107] "From this account it is clear that he only employed two causes: that of the essence [forms] and the material cause."[108] Correspondingly, in the passage from the *Physics* (above), Aristotle criticizes Timaeus' account on this same ground. The problem with the earlier account is the lack of clarity in the language, the mixing up of terms that results from not differentiating sufficiently. Again, according to Aristotle, the difficulty (*chalepon*) in understanding the *chora* arises out of understanding the *genos* of place in terms of the material cause of entities, a cause that indeed calls for the understanding of place as "room-occupying" or as *chora.* (Aristotle reads the meaning of *chora* literally, one translation of which can be "room.") In short, as Aristotle points out at the end of the passage above, what Timaeus has failed to see in not following a thorough analysis is that entities are different from their place.[109]

Aristotle's critique, worked out in concrete terms, goes even further and addresses the ambiguity caused by Timaeus' claim that *chora* is a third kind or form. Timaeus' account calls *chora* one of the three originary forms of all determinations of being. This indicates that spatiality has a role in the determination of beings. Aristotle rejects this association of spatiality with the determination of beings on the ground that place, as well as the indeterminacy and nonpresence figured by Timaeus' *chora,*

106. *Physics,* IV: 2, 209b13.
107. *Metaphysics,* I: 6, 987b20–988a14.
108. *Metaphysics,* I: 6, 988a10.
109. See note 79.

cannot be any of the four causes. "Again, how are we to suppose that place affects or determines things in any way? For it cannot be brought under any one of the four causal or essential determinants: not as the material of things, for nothing is composed of it; nor as their form or constituent definition; nor as their contemplated end [*telos*]; nor as setting them in motion [*kinesis*] or otherwise changing them."[110] Timaeus' figure is not a form but a phenomenon that must be engaged in its own event and in terms of presence. Here one finds a certain ambiguity in Aristotle's thought. Although Aristotle calls for a separation of the question of spatiality from that of things, so that space can be thought as marked by such a distinction, he continues to understand spatiality only in terms of presence and the systematic differentiation of its elements in language, i.e., in terms of a discourse still grounded on objective and ideal presence.

Traces of *Chora*: Aristotle's Withdrawal from the Question of Being

In light of the discussion of Timaeus' likely story and its figure of *chora,* Aristotle's discussion of place in his *Physics* appears as a transformative appropriation of Timaeus' figure. This is a change from an engagement of the events of beings, which intimates their essential alterity as well as that of the philosophical logos, to a discourse focused only on objective and ideal presence. This change occurs as Aristotle moves away from Timaeus' and Hesiod's *logoi,* their likely stories or accounts about beings, and develops a *logos apophantikos* that attempts to render all events intelligible through the rigorous analysis both of objective phenomena and of how these events of beings are engaged by the logos, or how all beings are said to be. Aristotle's project is to a certain extent phenomenological, in that he begins by looking at the phenomena. However, this phenomenological direction is, on the one hand, determined by a question of objective presence, the *ti estin* question; on the other hand, it is directed by the metaphysical aim of engaging the ever-present, unchanging governing causes and first principles behind all changing appearances. How does this logos engage the alterity of beings and the philosophical logos figured by Timaeus' *chora*? The reply to this question is difficult, since it involves Aristotle's withdrawal from Timaeus' intimations of alterity. At the same time, this withdrawal in the name of the logic of presence that

110. *Physics,* IV: 1, 209a20–24.

Aristotle's thought enacts intimates the alterity of the philosophical logos, as his account pushes further toward presence and remains therefore further from the arising to presence of events of beings. Moreover, the withdrawal of Aristotle's *logos apophantikos* is an aspect of that very logos in its event, and, in light of its withdrawal from the alterity of beings, therein appears the difficulty of engaging the alterity of that logos of presence in its withdrawing aspect, or alterity. In other words, something of the event of Aristotle's *logos apophantikos* remains to be engaged, and this aspect of its event at least calls for another way of understanding the philosophical logos, i.e., a way that will engage its alterity or withdrawal. These dense observations will require further development.

Aristotle's *logos apophantikos* moves away from a story-telling about beings, but in doing so it does not abandon or overcome what is essential to both Timaeus' account and Hesiod's poem: in all three cases the logos is given direction by the interpretation of events of beings in terms of objective and ideal presence. Furthermore, in the transfiguration of *chora* into *topos,* the difficulty figured by Timaeus' conception is lost, both in the sense that it is concealed and that Aristotle never addresses it. What is lost in Aristotle's discussion is the indeterminacy and non-presence of *chora,* together with the impossibility of "representing it," difficulties that figure the alterity of events of beings and of the philosophical logos. Aristotle's *logos apophantikos* turns these difficulties into logical mistakes, a criticism ultimately made on the basis of presence alone. In Aristotle's terms the difficulties of Timaeus' figure are perhaps overcome. However, such a view covers over the alterity of events of beings, and with it, the withdrawal (*choreo*) essentially operative in the presencing of beings. Furthermore, this covering over also indicates that the alterity figured by *chora* remains beyond Aristotle's discussion. Not only has the difficulty and alterity figured by Timaeus' story vanished from the questions both of being and of the logos, but at the same time this vanishing figures a certain withdrawal of Aristotle's *logos apophantikos* from the question of being. Because Aristotle does not take up the difficulties marked by *chora* in Timaeus' account, he never fully engages the question of being in the essential difficulty of its manifestations. Something happens in Aristotle's logos. In the powerful drive and with the full force of his *logos apophantikos* the question of being still seems to remain to a certain extent beyond Aristotle's thought. How is one to understand this withdrawal of Aristotle's philosophical logos from the question of being?

As in the case of Plato's *Timaeus,* in Aristotle's discussion one finds a journey to the limit of the sensible-intelligible world; this is what is in question throughout Aristotle's works under the rubric of a physics and a metaphysics.[111] However, his emphasis on objective and ideal presence will exercise a force that, in the drive for rendering all beings intelligible—visible, present—will at the same time exclude the alterity essential to events of beings and to their articulation. This exclusion does not simply mark a limit of language in regard to being. Rather, in the withdrawal of the question of being from Aristotle's powerful logos, the alterity of the philosophical logos is intimated. This intimation takes the form of an unavoidable question: Is the exclusion or vanishing of alterity, and is the forgetting of the withdrawal of being that occurs in the attempt to render all beings intelligible, an essential aspect of the philosophical logos? The exclusion of alterity does not merely mark a limit of language but also, through this limit, introduces the question of the essence of the philosophical logos, or the question of how the withdrawing of being and the exclusion of alterity is operative in the event of the philosophical logos. In other words, the limit one finds in the slipping of the withdrawing operative in manifestations of beings from representational logos as figured by *chora* does not enclose language, leaving it to "itself," but instead indicates the opening of language and of the philosophical logos to a difficulty that goes well beyond their traditional interpretation as mimetic representational tools subject to the necessities or logic of objective and ideal presence.

With these questions a further difficulty occurs: if a certain withdrawal is essential to the philosophical logos, then thought will be called to engage its alterity, i.e., to think both out of and in the event of the withdrawing operative in its events. If this is the case, the alterity of events of beings will only be engaged by a discourse that enacts this alterity, a discourse that will assert being in the awareness and enactment of a certain loss, slipping, withdrawal, alterity. In the exclusion of alterity by the forceful interpretation of events of beings in terms of presence Aristotle's philosophical discourse enacts a certain alterity operative throughout philosophical discourse, an aspect of the logos that up to this point is intimated twice, once by Timaeus' difficult story and once by Aristotle's forceful interpretation of *chora* as *topos.*

111. As previously indicated, Aristotle always asks whether a word articulates something, how that something may be engaged as what it is, and what is its activity, all questions that begin out of the limit of opinion and the standing tradition concerning the senses of being.

This series of issues appear in light of a tension between *muthos* (poetry) and *logos* (philosophical discourse). Aristotle begins his discussion of place out of two stories about beings, a curious beginning that echoes at the end something essential to Aristotle's *logos apophantikos* and hidden from it. This doubling of the logos seems to serve as a backdrop for the question of the philosophical logos in its alterity, and is felt throughout Plato's dialogues as well as in Aristotle's works, as Aristotle places the form of the dialogue under poetic/dramatic works, a determination that places his *logos apophantikos* once again apart from Plato's dialogues.[112] But perhaps, even in this constant differentiation the philosophical logos already intimates its alterity. On the one hand, we are speaking of a covering over of a withdrawal and indeterminacy essential to thought that is sustained by the assertive forms of the logos when understood only as a representational tool. On the other hand, in the striking attempt to understand *chora* in terms of a pure language of presence, thought remains under the stark shadow of its own inability to arrive at a representation of its events, a chiaroscuro apparent in the tension between *logos* and *muthos.*

Conclusion

This discussion has engaged two figures of spatiality essential to the question of being for Plato and Aristotle. Their close analysis uncovers a question of alterity at the heart of the ancients' battle for the sense of being, a question that appears as a double withdrawal. On the one hand, the indeterminacy of *chora* indicates something outside objective and ideal presence essential to the presencing of events of beings. On the other hand, this same withdrawal indicates the impossibility of representing or making intelligible the occurrences of beings. With respect to the philosophical logos, this last difficulty is felt in its extreme form in

112. See, for example, Plato's *Republic* and his discussion of poetry in the city, in which the separation of the philosophical logos from poetry is presented with such sustained concentration and intensity that the philosophical logos cannot be thought separate from the force of the poet's accounts. Thus, in the entanglement of the two and in the affirmation of the power of poetry/accounts, the philosophical logos remains inseparable (although differentiated) from accounts about beings.

In the first page of the *Poetics,* Aristotle identifies the dialogue as a form of poetry/drama.

Aristotle's discussion of place (*topos*). In his discussion Aristotle attempts to give an account of the spatiality of events of beings based only on objective and ideal presence and the logic of this presence. Ultimately, Aristotle's logos fails to address the difficulties figured by *chora.* This failure enacts a withdrawal operative in the very event of Aristotle's logos: Aristotle's *logos apophantikos* does not engage the events of beings and thought in the full difficulty of their spatiality, i.e., in their withdrawal or alterity. Therefore, not only do all events of beings remain beyond this logos, but also the very event of this logos in its withdrawing aspect will remain unaccountable in its own terms of presence alone. These difficulties with the question of being as engaged by Plato and Aristotle remain to be engaged by Heidegger in his rekindling of the question of being in *Being and Time.* In terms of Heidegger's focus on temporality as the ground of the occurrences of beings, one can say that this look back to Plato and Aristotle recalls *Being and Time,* as well as its readers, to issues of spatiality and alterity, difficulties that seem to be overshadowed by Heidegger's swift move in the first page of *Being and Time* from the Greek quotation to the question of temporality. In this sense, it will be to a certain extent by way of a transgression and interruption of *Being and Time*'s focused discourse on temporality as the origin of beings that alterity and spatiality will be engaged, a transgression and interruption that intimates something foreign and yet operative in the question of being.[113]

113. This transgression of the discourse's focus on temporality in *Being and Time* does not seem altogether outside the path of thinking of the tradition. Is it not in light of the interruption of the philosophical logos that Heidegger's *Being and Time* begins? Is it not as a figure of transgression and interruption that Timaeus' third kind appears? Is it not by transgression that Aristotle will take *chora* as *topos*? Something like interruption and transgression, something always foreign and yet of the logos (no longer in the genitive sense of correspondence, or even in a resolvable dialectic) will be operative in the weaving of the philosophical logos and in its determinations of all senses of beings.

2

Exilic Thoughts

Alterity and Spatiality in the Project of *Being and Time*

Introduction

Throughout the development of Heidegger's thought spatiality becomes increasingly present in the articulation of the question of being. A change in focus seems to occur in Heidegger's thinking from his early single emphasis on the temporal horizon of being to his later preoccupation with the spatiality of beings.[1] In *Being and Time* the discussion of being concerns mainly its temporality. Heidegger states in the first page of

1. There are few works on Heidegger and spatiality. The two main works I have followed are Friedrich-Wilhelm von Herrmann, "Wahrheit-Zeit-Raum," in *Die Frage Nach der Wahrheit* (Frankfurt am Main: V. Klostermann, 1997); and Edward Casey, "Proceeding to Place by Indirection," in *The Fate of Place* (Berkeley and Los Angeles: University of California Press, 1997). I have also referred to Emil Kettering, *Nähe: Das Denken Martin Heideggers* (Pfüllingen: Neske, 1987); Didier Franck, *Heidegger et le problème de l'espace* (Paris: Minuit, 1986); and Maria Villa-Petit, "Heidegger's Conception of Space," in *Martin Heidegger: Critical Assessments,* ed. Christopher Macann (New York: Routledge, 1992).

Being and Time that his project is "the interpretation of *time* as the possible horizon for any understanding whatsoever of being."[2] In the 1935–36 lectures that formed the basis for "The Origin of the Work of Art," being and spatiality appear inseparably intertwined in the discussion of the site of the Greek temple.[3] Then, in *Contributions to Philosophy* (1938), Heidegger discusses originary temporality (*ursprüngliche Zeit*) in terms of the "time-space" (*Zeit-Raum*) of the truth of being, where spatiality assumes equiprimordial relevance with temporality in the disclosedness of being.[4] Finally, in the 1950s in "Art and Space" (1969), Heidegger discusses the disclosedness of being in terms of "place" (*Ort*).[5]

The issue of spatiality in Heidegger's thought leads to the alterity of events of beings and thought, an aspect of the question of being that points toward the possibility of a thinking (a philosophical discourse) that engages alterity outside the interpretation of beings in terms of objective and ideal presence. This chapter shows that the difficulty of engaging the alterity of events of beings and thought is tied to Heidegger's engagement of spatiality in *Being and Time.* This is done by first showing that spatiality appears as a figure of alterity in *Being and Time,* and by then turning to the way Heidegger's project in general takes up the issue of alterity through his development of an "apophantic logos." This examination ultimately shows that Heidegger's project opens a

2. *BT,* 1, italics in original (*SZ,* 1).

3. "Der Ursprung des Kunstwerkes," *in Holzwege, GA* 5. Cf. Heidegger's 1931–32 version of the lecture "Vom Ursprung des Kunstwerks. Erste Ausarbeitung," in *Heidegger Studies* 5 (1989): 5–22. Compare how this same relation of spatiality and being appears in "Building, Dwelling, Thinking," in *Basic Writings,* ed. David Farrell Krell (San Francisco: Harper, 1993). See also "Bauen, Wohnen, Denken," in *Vorträge und Aufsätze* (Pfüllingen: Neske, 1985).

4. In *Contributions to Philosophy,* tr. Parvis Emad and Kenneth Maly (Indianapolis: Indiana University Press, 1999), *Zeit-Raum* occurs as the *Ab-grund,* the abysmal truth of being. See also especially "Der Zeit-Raum als der Ab-grund," in *Beiträge zur Philosophie (Vom Ereignis), GA* 65, 371–79.

My interpretation of Heidegger's thinking in the 1930s is mainly based on the work of John Sallis, of Charles Scott, and of Daniela Vallega-Neu's *Die Notwendigkeit der Gründung im Zeitalter der Dekonstruktion* (Berlin: Duncker & Humblot, 1997); as well as on the work of the contributors to *Companion to Heidegger's 'Contributions to Philosophy,'* ed. Charles Scott, Susan Schoenbohm, Daniela Vallega-Neu, and Alejandro Vallega (Indianapolis: Indiana University Press, 2001).

5. "Place (*Ort*) always opens a region (*Gegend*) by gathering things in this region into their belonging together. . . . In the place the gathering plays in the sense of the releasing-sheltering of the things into their region. [Der Ort öffnet jeweils eine Gegend, indem er die Dinge auf das Zusammengehören in ihr versammelt. Im Ort spielt das Versammeln im Sinne des freigebenden Bergens der Dinge in ihre Gegend]." "Die Kunst und der Raum," in *Aus der Erfahrung des Denkens, GA* 13, 207, my translation.

space of thought that, because of its awareness of its very alterity, remains open to the occurrences of beings in their diversity and withdrawal. As we saw in the previous chapter, these are aspects of the events of beings that have been covered over by traditional metaphysical and transcendental interpretations of being in terms of concepts, principles, ideas, and teleology grounded on objective and ideal presence, as well as by the interpretation of the philosophical discourse as a mimetic tool charged with the task of making the unchanging Being of beings intelligible.

This encounter with Heidegger's project rigorously traces the motion of his thought, a task that calls for utmost attentiveness to both the content (what is said) and the structure of the work (the structure of the disclosedness of being as worked out by Heidegger's analysis of dasein's fundamental temporality). Furthermore, this also requires that one meticulously follow and persist in the engagement of Heidegger's thought, by listening and marking how it occurs in those exclusions, arrests, pauses, and silences constitutive of that thought's event. Given the character of these issues, listening for them in Heidegger's discourse will be an obvious requirement for engaging the alterity and the exilic dimensions of events of beings and of thought in his work. Throughout this book I will refer to these constitutive aspects of Heidegger's thought as performative.[6]

Spatiality and the Question of Being

When one looks at *Being and Time* there seems to be little place for the issue of spatiality in Heidegger's revival of the question of being. The central focus of *Being and Time* is temporality, and not spatiality. The book as a whole embodies Heidegger's particular concern with temporality. Only three sections in the book discuss spatiality as their theme (§22–24).[7] In a later section (§70), Heidegger attempts to show how the spatiality discussed in the previous sections arises out of temporality.[8]

6. The term "enactment" refers to Heidegger's later thought, for example *Contributions to Philosophy,* where he understands the very event of his thought as the enactment of the disclosedness of beings in their epochal possibility and situatedness. The German term is *Vollzug.* This term is therefore synonymous with the enactment of the disclosedness of being in events of thought, and does not point outside or beyond the presencing of beings as such.

7. *BT,* 94–105 (*SZ,* 101–13).

8. *BT,* 335–38 (*SZ,* 367–69).

However, from the book's beginning, spatiality makes itself heard not only as part of the thinking of the question of being, but also as a powerful aspect of that question, one that will even endanger Heidegger's project. As it will become apparent, it is precisely the exclusion or suspension of spatiality from the articulation of the question of being that indicates its powerful role in Heidegger's thought.

Despite the primacy of temporality in *Being and Time,* Heidegger begins his analysis of dasein from an analysis of spatiality, and not from temporality. This is clearly seen when one looks at the general structure of Heidegger's analysis of dasein. In *Being and Time* the fundamental point of departure for the analysis of the meaning of the question of being in its temporal disclosedness is the ontological difference between beings (*Seinden,* entities at hand) and being (*Sein*). In light of this ontological difference the disclosedness of being in its temporality is sought through the analysis of "dasein" in its ontico-ontological way of being.[9] *Being and Time* is thus divided into two parts. In the first Heidegger gives a fundamental analysis (*Fundamentalanalyse des Daseins*) of dasein's ontological structure, or its being-in-the-world (*In-der-Welt-sein*), in order to give the conceptual structure of the preconceptual disclosedness of being in dasein's being-in-the-world. In the second part, as its title *Dasein und Zeitlichkeit* indicates, Heidegger reinterprets this structure in terms of fundamental temporality.

In Part I, §12 of *Being and Time,* Heidegger first orients his analysis of dasein by establishing the ontological difference between dasein and entities at hand. In this section, for the sake of analysis, Heidegger breaks up the unified ontological structure of dasein, or being-in-the-world (*In-der-Welt-sein*), into three components.[10] These are: "world" (*in der Welt*) and the idea of "worldhood" (*Weltlichkeit*); the "entity" that in every case has being-in-the-world as the way in which it is (*das Seiende, das je in der Weise des In-der-Welt-seins ist*); and "being-in" (*In-Sein*) as such.[11] Then, the analysis of being-in-the-world is oriented from a preliminary discussion of the third element "being-in" (*In-Sein*).

9. It should be added that this analysis is also to be understood as the analysis of the question of being in its disclosedness.

10. *BT,* 50 (*SZ,* 53).

11. Although in English *In-sein* is translated as "Being-in," the order of the words in German is meant to mark the ontological difference between the spatiality of things at hand, or "being in something" (*sein in . . .*), and the spatiality of dasein, or *In-sein.* Although not grammatically correct in English, it may be helpful to think of dasein's *In-sein* literally as "in-being." This expression not only marks the difference Heidegger accentuates in his language, but also keeps

> Before making these three phenomena the themes for special analysis, we shall attempt by way of orientation to characterize the third of these factors.
>
> What is the meaning of *Being-in* [*In-Sein*]? Our proximal reaction is to round out this expression to "Being-in 'in the world,'" and we are inclined to understand this Being-in as "Being in something" [*Sein in* . . .]. This latter term designates the kind of Being which an entity has when it is "in" another one, as the water is "in" the glass, or the garment is "in" the cupboard. By this "in" we mean the relationship of Being which two entities extended "in" space have to each other with regard to their location in that space. . . . All entities whose Being "in" one another can thus be described have the same kind of being—that of being present at hand—as things occurring "within" the world. Being-present-at-hand [*Vorhandensein*] "in" something which is likewise present at hand, and being-present-at hand-along-with [*Mitvorhandensein*] in the sense of a definite location-relationship with something else which has the same kind of Being, are ontological characteristics which we call *categorial*: they are of such sort as to belong to entities whose kind of Being is not of the character of dasein.
>
> Being-in [*In-Sein*], on the other hand, is a state of Dasein's Being; it is an existentiale. So one cannot think of it as the Being-present-at-hand of some corporeal thing (such as a human body) "in" an entity which is present-at-hand. Nor does the term "Being-in" mean a spatial "in-one-another-ness" ("*In*"-*einandersein*) of things present-at-hand, any more than the word "in" primordially signifies a spatial relationship of this kind.[12]

Perhaps a striking aspect of this discussion is that, with the exception of the last paragraph, it is a direct paraphrasing of Aristotle's discussion of place in the *Physics,* Book IV. But this is only an indication of the more powerfully telling aspect of the discussion, particularly with respect to

us close to Heidegger's understanding of dasein as a being that occurs always with a world; being always in a world is indicated by "in-being." At the same time, it would be a mistake to dissociate Heidegger's term from the everyday uses of language, since this suggests a complete separation between the language of everydayness and the language of philosophy. In other words, in using a term like "in-being" to translate Heidegger's word one runs the risk of turning Heidegger's German into a technical language that is severed from its ordinary usage, thus losing the range of lived sedimentation and openness that language itself offers in thinking.

12. *BT,* 79, M&R translation (*SZ,* 53–54).

the issue of spatiality in *Being and Time.* The primary aim of this passage is to make a differentiation between the being of entities at hand and the occurrence or way of being that is "dasein" (being-t/here). This is the crucial differentiation that indicates the ontological difference between beings and the being of dasein.[13] By bringing forth dasein's unique way of being, this distinction opens a path toward the disclosedness of being in its temporal horizon. In other words, being will be engaged in its temporality on the basis of this differentiation.[14] But what is most surprising about the differentiation is this: Although the aim of the passage is to point toward the fundamental temporality of dasein, Heidegger makes his overture toward the question of being out of a discussion of spatiality. The ontological difference is first uncovered through a spatial differentiation between the objective spatiality of things—for example, in terms of "being in something" (*sein in*)—and spatiality as a mode of being, a certain "being-t/here" or "dasein," which Heidegger indicates with the term *In-Sein* (literally "in-being"), in contrast to objective being. Heidegger's beginning from spatiality certainly indicates that this issue is operative with regard to the disclosedness of the question of being understood as dasein. But to what extent is spatiality operative with regard to the question of being?

The cited passage does not only introduce spatiality but points to a radicalization of this issue as well. On the one hand, "being in something" (*sein in etwas*) refers to the "space" of things objectively present at hand, as well as to the concepts that ground such interpretation (that is, to the "space" of nature, the physical sciences, and the primary philosophical interpretations of spatiality as the "vessel," or container, of entities). On the other hand, Heidegger then goes on to speak of dasein's *In-Sein* as "dwelling" (*wohnen*),[15] a modality of being he will then ultimately ground on "care" (*Sorge*), the temporal modality that holds

13. The differentiation Heidegger is making is not a repetition of the metaphysical dualism between unchanging being (traditional ontology) and changing beings (as objective beings or entities at hand). Dasein is ontico-ontological, and this is not a category that will accept the metaphysical separation.

14. Although the "ontological difference" is not yet explicitly thematized in *Being and Time,* it is clear from the opening differentiation between entities and dasein's ontico-ontological way of being that this difference occurs as an operative element of Heidegger's thought in *Being and Time.* Heidegger speaks of the ontological difference at about the same time as the publication of *Being and Time* (1927) in his Marburg lectures from the Summer semester of 1927. See *Die Grundproblem der Phänomenologie,* Part II, Chapter 1 (*GA* 24, 473). As von Herrmann points out in his editorial comments, this text concerns the fundamental traits of the question of being.

15. *BT,* 51 (*SZ,* 54).

together the structure of being-in-the-world. This indicates that spatiality is to be thought in terms other than its traditional interpretations as objective and ideal presence. Spatiality must be rethought in terms of dasein's temporality and finitude.[16] Later in the same section Heidegger explains that, although the spatiality of entities at hand is not dasein's way of being, this does not mean that there is not a characteristic spatiality that belongs to it: "In the first instance it is enough to see the ontological difference between Being-in as an *existentiale* and the category of the 'insideness' which things present-at-hand can have with regard to one another. By thus delimiting Being-in, we are not denying every kind of 'spatiality' to Dasein. On the contrary, Dasein itself has a 'Being-in-space' of its own; but this in turn is possible only *on the basis of being-in-the-world in general.*"[17] Dasein always occurs spatially. In order to think dasein's being-in-the-world, as well as engage the question of being, it will be necessary not only to take up dasein's temporality but also to engage the issue of spatiality. This is because spatiality is not less an issue for dasein than is temporality.

That spatiality is for Heidegger a necessary and indispensable problem in the engagement of the question of being as disclosed in dasein is clear in the following passage:

> The perplexity still present today with regard to the interpretation of the being of space is grounded not so much in an inadequate knowledge of the factual constitution of space itself as in the lack of a fundamental transparency of the possibilities of being in general and of their ontologically conceived interpretation. What is decisive for the understanding of the ontological problem of space lies in freeing the question of the being of space from the narrowness of the accidentally available and, moreover, undifferentiated concepts of being, and, with respect to the phenomenon itself, in moving the problematic of the being of space and the various phenomenal spatialities in the direction of clarifying the possibilities of being in general.[18]

16. The terms "temporality" and "finitude" are ultimately synonymous in *Being and Time* as far as the analysis of dasein's temporality is concerned.

17. *BT,* 82, M&R translation, italics in original (*SZ,* 56).

18. *BT,* 104–5 (*SZ,* 113).

In this passage spatiality and being appear inseparably entangled as they do perhaps in no other place in *Being and Time.* First of all, it indicates that one can engage dasein's spatiality or engage the issue of spatiality in general only by thinking outside the metaphysical tradition of objective and ideal presence. Indeed, spatiality will have to be turned toward the question of being. This turn is not a matter of waiting for the analysis of temporality in order to then think spatiality, an impression given by Heidegger's emphasis on temporality. Rather, it will now be a matter of thinking spatiality along with the question of being, i.e., of engaging the question of being in its spatiality at each step of the analysis of dasein. Since dasein is fundamentally spatial, spatiality can only be understood in conjunction with the question of being. This is clearly indicated by the fact that Heidegger discusses spatiality in both Parts I and II of *Being and Time,* and the passage above indicates this turn to spatiality in dasein as well. It begins by recognizing the difficulty of engaging spatiality in traditional terms, but then goes on to connect this difficulty to the question of being. This second aspect of Heidegger's discussion has almost imperceptibly translated the issue of spatiality from a question of objective and ideal presence to one aspect of the question of being that is inseparable from it and always operative in events of beings as found through the analysis of dasein (being-t/here).

As the last paragraph of this quotation also indicates, for Heidegger there is a certain spatiality that always belongs to dasein. Inasmuch as this is the case, a figure of spatiality can be said to appear at the very point of engagement of the occurrences of beings through the analysis of the question of being in the modality of dasein (being-t/here). In this there is a distinct inseparability of events of beings and spatiality. According to Heidegger the question of being occurs in dasein, and dasein may only occur as a being-in-the-world along with others, a being with others that makes room for the being of entities at hand. As Heidegger indicates in §3, "When we let entities within-the-world be encountered in the way which is constitutive for Being-in-the-world, we 'give them space' [*Raum-geben*]. This 'giving space,' which we also call '*making room*' [*Einräumen*] for them, consists in freeing the ready-to-hand for its spatiality."[19] This means that spatiality is not only inseparable from the question of being, but that it plays a fundamental role in the occurrences of beings. In this passage dasein's spatiality echoes the disclosive recep-

19. *BT,* 146, M&R translation, italics in original (*SZ,* 111).

tivity found in Timaeus' figure of spatiality, *chora.* In its opening of "spaces" for beings in the coming forth of beings, dasein's spatiality in being-in-the-world suggests something akin to the freeing receptivity of the *chora.* The question that arises now is: To what extent does dasein's spatiality figure the difficulties of Timaeus' *chora*?

Dasein's Spatiality as a Figure of Alterity

When one looks closely at Heidegger's first engagement with spatiality in *Being and Time,* it becomes apparent that spatiality is not only difficult for Heidegger but that it appears as a figure of alterity as well. Spatiality is introduced by an interruption of Heidegger's discourse on temporality, and at the very beginning of his analysis of dasein's temporality. Heidegger must abruptly interrupt his discourse on dasein's everydayness in order to indicate the necessary suspension of spatiality from his analysis. It will be, then, by way of a performative interruption that spatiality will be introduced to Heidegger's discussion; an introduction that engages spatiality as a figure of alterity both with respect to Heidegger's main discourse and intention and with regard to what is to come (the analysis of spatiality out of dasein's essential temporality).

In Part I, Chapter 3, Heidegger develops an analysis of the "worldhood of the world" (*Die Weltlichkeit der Welt*). His aim is to start with dasein's most proximate dealings with the world, i.e., the environment (*Umwelt*), as a way toward dasein's fundamental temporality. In setting out toward temporality Heidegger must interrupt his discussion in order to broach a certain difficulty with the spatial aspect of his analysis. As he introduces this section Heidegger pauses to note that the spatial character suggested by the word "environment" (*Umwelt*) does not indicate that it is primarily a spatial phenomena. "The expression 'environment' [*Umwelt*] contains in the 'environ' [*um*] the suggestion of spatiality. Yet the 'aroundness' [*Umherum*] which is constitutive for the environment does not have a primarily 'spatial' meaning."[20] This passage seems to point merely to the separation of spatiality from dasein's *In-Sein* as discussed above. However, the few sentences are a condensation of a whole paragraph in *The Prolegomena to the History of the Concept of Time*

20. *BT,* 94, M&R translation (*SZ,* 66).

(1925), an earlier draft of *Being and Time* in lecture form.[21] When one looks at this earlier and longer version of the sentence in *Being and Time,* a certain difficulty concerning spatiality begins to appear, and with it the power of spatiality in the engagement of the question of being.

In §21b of the *Prolegomena* Heidegger states why dasein's being-in-the-world and its *In-Sein* modality should not be understood in terms of space.

> The "*around*" and the "*aroundness*" are not to be taken primarily spatially, and not spatially at all if spatiality is defined in terms of the dimensionality of metric space, the space of geometry. On the other hand, however, the continual resistance to spatiality which we are forced to adopt in the determination of in-being, in the characterization of world and still more in the account of the environing world, the constant necessity here to suspend a specific sense of spatiality, suggests that in all of these phenomena a certain sense of something like spatiality is still in play. This in fact is the case. For just this reason it is important from the start not to miss the question of the structure of this spatiality, that is, *not* to start from the spatiality which is specifically geometrical, a spatiality which is discovered in and extracted from the primary and originary space of the world. Since it is a question of understanding the primary sense of world, a particular idea of space understood in terms of metric space must first be put out of play. On the contrary, we shall learn to comprehend the sense of metric space and the particular modification which motivates metrics in spatiality by reference to a more originary spatiality [*ursprünglichere Räumlichkeit*].[22]

In order to understand the structure of the world a certain interpretation of spatiality must be put out of play. Heidegger is referring to the interpretation of spatiality in terms of objective and ideal presence (the interpretation we find in the "space" of geometry, and the concepts of "space"

21. The page numbers will refer to *GA* 20; the English translation, unless noted, comes from *History of the Concept of Time,* tr. Theodore Kisiel (Indianapolis: Indiana University Press, 1992). This is the manuscript of Heidegger's lectures in Marburg for the summer semester of 1925.

22. *GA* 20, 230. I translate *ursprünglichere Räumlichkeit* as "originary spatiality" in order to distinguish Heidegger's term *ursprünglichere* from "original." Heidegger's use of this word refers to the temporality of events of beings and not to "original" phenomena in the sense of unchanging or transcendental structures or categories of being.

associated with it, namely, the concepts of spatiality associated with "nature," the physical sciences, and philosophical categories). However, as the last sentence of the paragraph indicates, this does not call for a mere refutation of metaphysics and transcendental philosophy. The point is to gain a way of approaching spatiality, a way of moving from traditional concepts of "space" to a "more originary spatiality." *In-Sein* does not refer to ideal or objective space. Therefore, the traditional approaches to spatiality will not suffice. In this paragraph one finds a double suspension of spatiality based on Heidegger's analysis of the question of being. On the one hand, the natural attitude toward spatiality as objective, and its attendant conceptual categories, must be set aside. On the other hand, the paragraph points to the issue of a more originary spatiality, i.e., dasein's spatiality or the spatiality of the appearing of events of beings. However, this spatiality is ultimately suspended as Heidegger moves to uncover the fundamental temporality of dasein, and through this analysis, the temporality fundamental to the occurrences of beings.

Because of the necessary suspension of the objective and ideal interpretations of spatiality the issue of a more originary spatiality must also remain under silence. Originary spatiality cannot be engaged in terms of nature, categories, or presence in general. The originary spatiality of events of beings must therefore wait, even in spite of the operative role of spatiality in the very engagement of the question of being (a point that is always illustrated by Heidegger's return to issues of spatiality throughout his book). But why must spatiality wait? What does this necessary suspension of spatiality indicate about it?

Heidegger's call for the necessary suspension of spatiality in the engagement of the question of being indicates spatiality's powerful role and irreducible effect on his project. It is the danger of failing to engage the question of being by repeating the metaphysical interpretation of the occurrences of beings and particularly of dasein (as a subject, as a rational animal, an entity among other entities "in" a world) that ultimately requires Heidegger's silencing of spatiality as he sets out in his project. If the spatiality of being is taken up again in terms of presence, then the whole analysis of the question of being is reinscribed into the conceptual apparatus of metaphysics and objective and ideal presence. In other words, spatiality presents a danger to the whole project of *Being and Time.* Engaging dasein's spatiality in terms of presence would mean failing to engage the question of being. This is why Heidegger must keep silent concerning spatiality, although it is inseparable from the question of being.

In his introduction to and suspension of spatiality Heidegger opens his thinking to a difficulty outside his discourse on temporality. His silence is necessary because of the intrinsic play of spatiality in the question of being, and because of the danger of failing to engage the question of being by beginning from a discussion of its spatiality, which, in treating this figure in terms of presence, will have already reinscribed the question of being into the metaphysical/transcendental tradition. Spatiality figures a double silence: on the one hand, as what must remain outside Heidegger's discussion, and on the other hand, as what is yet to be thought. Thus, it is as such a figure of alterity—as that which remains outside of Heidegger's analysis—that the spatiality of the question of being will have been first introduced by Heidegger's thought in *Being and Time.*

That spatiality appears as a figure of alterity in *Being and Time* recalls Timaeus' *chora* in Plato's *Timaeus,* and in doing so also brings forth the difficulties found in this figure. The two main difficulties are the indeterminacy and withdrawing character of *chora,* and the issue of a logos that may engage this figure of the nonpresence or withdrawal at play in events of beings. The remaining part of this chapter discusses how Heidegger's project aims to engage these difficulties.

Language and the Question of Being

The question of language is inseparable from and fundamental to the question of being throughout Heidegger's career. Heidegger himself indicates this in "A Dialogue On Language" (published in *On the Way to Language* [*Unterwegs zur Sprache*], 1959). In this late piece Heidegger looks back on his path and says that the one ever-present and most difficult problem in his thinking is the question of language and being.

> JAPANESE: [. . .] Again and again it was said that your question circled around the problem of language and Being.
>
> INQUIRER: In fact, this was not too difficult to discern; for as early as 1915, in the title of my dissertation "Dun Scotus' Doctrine of Categories and Theory of Meaning," the two perspectives came into view: "doctrine of categories" is the usual name of the discussion of the Being of beings; "theory of meaning"

means the grammatica speculativa, the metaphysical reflection on language in its relation to Being. But all these relationships were then still unclear to me.

JAPANESE: Which is why you kept silent for twelve years.

INQUIRER: And I dedicated *Being and Time*, which appeared in 1927, to Husserl, because phenomenology presented us with possibilities of a way.

JAPANESE: Still, it seems to me that the fundamental theme, "Language and Being," stayed there in the background [. . .]. [. . .] The question of language and Being is perhaps a gift of that light ray which fell on you.Inquirer: Who would have the audacity to claim that such a gift has come to him? I only know one thing: because reflection on language, and on being, has determined my path of thinking from early on, therefore their discussion has stayed as far as possible in the background. The fundamental flaw of the book *Being and Time* is perhaps that I ventured forth too far too early.

JAPANESE: That can hardly be said of your thought on language.

INQUIRER: True, less so, for it was all of twenty years after my doctoral dissertation that I dared discuss in class the question of language. [. . .] In the summer semester of 1934, I offered a lecture series under the title "Logic." In fact, however, it was a reflection on the logos, in which I was trying to find the nature of language. Yet it took nearly another ten years before I was able to say what I was thinking—the fitting word is still lacking even today. The prospect of the thinking that labors to answer to the nature of language is still veiled, in all its vastness.[23]

For Heidegger the question of being is always at the same time the problem of language. This is the case in *Being and Time,* in which his radicalization of Husserlian phenomenology occurs as a means of approaching this problem. The difficulty of language is so great that it always occupies and limits Heidegger's thinking. It requires his constant reflection. It "determines" his thinking, and yet, throughout his work, he says, he could not discuss it directly and so kept it in the background. Only after

23. "A Dialogue On Language," in *On the Way to Language,* tr. Peter D. Hertz (San Francisco: Harper, 1971), 6–8.

thirty years of reflection does Heidegger claim to be able to say what he thought, and he does this in full awareness of a lack for words. Finally, he finds only incompleteness in his saying.

One finds indications of this difficulty already in Heidegger's *Being and Time*, where he begins with a strong remark that exposes his project from the start to the difficulties of developing a philosophical discourse that will engage the occurrence of beings in their alterity. Heidegger asserts that "[t]he first philosophical step in understanding the problem of being consists in avoiding 'μυθον τινα διηγεισται' [*muthon tina diegeistai*] in not telling a story."[24] The quote in Greek comes again from Plato's *Sophist* (242c). The passage calls for not telling a story about beings, i.e., for not interpreting the events of beings in terms of objective and ideal presence. This observation certainly calls to mind Timaeus' story as well as Aristotle's *logos apophantikos*, two accounts about beings, two stories told in terms of objective and ideal presence. At the same time, Heidegger's words also call to mind the difficulties found in those accounts with respect to spatiality and alterity, and in doing this, they introduce these difficulties into his discourse in *Being and Time*. How does Heidegger's philosophical discourse engage the alterity or withdrawal of events of beings? How does Heidegger's project take up the even more difficult problem of engaging alterity by enacting a philosophical discourse that echoes the alterity operative in its event? Heidegger's words are not only a warning against metaphysics and a language of presence and representation, but also indicate the difficulty fundamental to his own attempt to rekindle the ancient battle for the question of being.

The most obvious and immediate problem raised in *Being and Time* concerning the question of being is that it has been forgotten, covered over by metaphysics.

> At the outset [§1] we showed that the question of the meaning of Being was not only unresolved, not only inadequately formulated, but in spite of all interest in "metaphysics" has even been forgotten. Greek ontology and its history, which through many twists and turns still define the conceptual character of philosophy today, are proof of the fact that Dasein understands itself and being in general in terms of the "world." The ontology which thus arises is ensnared by the tradition, which allows it to sink to the

24. *BT*, 5 (*SZ*, 6).

> level of the obvious and become mere material for reworking (as it was for Hegel). . . . In so far as certain domains of being become visible in the course of this history and henceforth chiefly dominated the range of problems (Descartes' *ego cogito*, subject, the "I," reason, spirit, person) the beings just cited remain unquestioned with respect to the being and structure of their being, which indicates the thorough neglect of the question of being.[25]

The philosophical tradition has understood being in terms of entities at hand, of things, and of their objective and ideal presence. In this sense being becomes a certain unchanging ground that gives identity and continuity to an otherwise ever-changing process of beings in their coming to be in passing away. Plato's *eidos,* Aristotle's *ti hen einai* [that which is something], Descartes' *substantia,* all implicitly hold to an analogous understanding of being in terms of the logical necessity for the identity of entities objectively present at hand.[26]

In the *Prolegomena to the History of the Concept of Time,* where Heidegger gives an explanation of how the covering over of being occurs, he shows how language and this concealment of being are inseparably linked:

> Thus, discourse and λογος [*logos*] for the Greeks assumes the function of theoretical discussion. The λογος accordingly gets the sense of exhibiting what is talked over in its whence and what about. The exhibition of the entity in its reasons, what is said, what is exhibited in discourse, the λεγομενον [*legomenon*] as λογος, is then the ground or reason, what is apprehended in understanding comprehension, the *rational.* Only in this derivative way does λογος get the sense of *reason*, just as *ratio*—the medieval term for λογος—has the sense *of discourse, reason, and ground. Discoursing about . . . is exhibiting reasons, founding, letting something be seen referentially in its whence and how.*[27]

A certain understanding of being articulated through a reductive language has covered over the question of being by reducing the logos to a discourse "about" beings and their "reason." In *Heidegger's Ways,* Hans-Georg Gadamer gives a helpful paraphrasing of Heidegger's thought:

25. *BT,* 19 (*SZ,* 21–22).
26. *BT,* 17–23 (*SZ,* introduction, 6); *BT,* 83–94 (*SZ,* I: 3, part B).
27. *GA* 20, 365, italics in original.

> What manifests itself as *eidos*, i.e., as an unchangeable determinateness showing the "What-of-Being" [*Was-Sein*], understands "Being" implicitly as a continuous presence [*Gegenwart*], and this determines as well the meaning of unconcealedness, that is, of truth, and establishes the criterion of right and wrong for every assertion about beings. The claim "Theatetus can fly" is false because people are incapable of flight. In this way, through his reinterpretation of the Eleatic doctrine of Being as the dialectic of Being and Non-being, Plato grounds the meaning of "knowledge" in the λογος which allows assertions about the beingness of beings, that is, about the What-of-Being.[28]

Gadamer's succinct sketch shows that, for Heidegger, the problem of the covering over of the question of being occurs inseparably with the setting up of an assertive logos that reduces being to presence by limiting the way the question of being is articulated. One finds this metaphysical understanding of the world most radically expressed in the dualistic view of being as the ever-passing and changing being of things in the world, and the unchanging and ever-present essence that secures an identity for these otherwise ephemeral things. These are two sides of the one coin that gives value to being by asking the question of being in terms of presence, in a language of "assertions about the beingness of beings, that is, about the What-of-Being."

In light of this covering over of the question of being by the language of metaphysics, Heidegger's project in *Being and Time* must occur as a break with the traditional articulation of being as the being of beings in their objective and ideal presence. As Heidegger indicates—as we have seen in the passage of *Being and Time* where he states that we can no longer tell a story about beings—this break concerns language:

> The Being of entities "is" not itself an entity. If we are to understand the problem of Being, our first philosophical step consists in not "μυθον τινα διηγεισθαι" [*Sophist*, 242c], in not "telling a story"—that is to say, in not defining entities as entities by tracing them back in their origin to some other entity, as if Being had the character of some possible entity. Being, as that which is asked about, must be exhibited in a way of its own, fundamentally dif-

28. Hans-Georg Gadamer, "Plato," in *Heidegger's Ways*, tr. John W. Stanley (Albany: SUNY Press, 1994). See also *Heideggers Wege* (Tübingen: Mohr, 1983).

> ferent from the way in which entities are discovered. Accordingly, *what is to be found out by asking*—the meaning of Being—also demands that it be conceived in a way of its own, fundamentally contrasting with the concepts in which entities acquire their determinate signification.[29]

Here again, Heidegger calls for a double task. On the one hand, thought must begin to uncover the question of being; on the other, this must occur through a philosophical discourse that does not repeat the interpretation of beings in terms of presence and representation. The change in the question figures a change in language, in philosophical discourse.[30] The covering over, as well as the possibility, of the engagement of the question of being is inseparably at play in the very saying of being, in the way the philosophical discourse will engage or fail to engage the question of being in its disclosedness.[31] This is what Gadamer's discussion indicates, and this is why Heidegger introduces the difficulty of language at the outset of *Being and Time.* At the same time this also indicates that the alterity of events of beings will be an issue always caught in manifestations of beings in language. Thus, the question arises: How is the alterity of beings engaged in Heidegger's discourse?

Heidegger's Project of an Apophantic Logos

The intrinsic role of language in the covering over of the question of being indicates a negative aspect of language. Indeed, both Timaeus' and Aristotle's accounts seem to serve as full examples of philosophical

29. *BT,* 26, M&R translation, italics in original (*SZ,* 6).

30. This also indicates a change in the understanding of language, since its mimetic function will no longer determine how one understands it. Furthermore, this also indicates a modification in the traditional juxtaposition of logos and *muthos*: Does the abandonment of stories about beings not release the *muthos* to its event by indicating at least the possibility of understanding it in terms that do not reduce it to representation alone? See Rodolphe Gasché's powerful engagement of the question of literature and philosophy in *The Tain of the Mirror* (Cambridge, Mass.: Harvard University Press, 1986).

31. Throughout this work I juxtapose objective and ideal presence and the presencing of beings in their withdrawal or alterity. This does not suggest the search for a pure language beyond presence. On the contrary, the juxtaposition is meant to engage presence outside of the unquestioning repetition of the interpretation of being as presence.

discourses that remain with presence in spite of their difficulties. At the same time, this language of presence has a power that seems to remain unshaken, even in the face of such figures of alterity as Timaeus' *chora.* Indeed, language in its occurrence has the power to conceal the disclosedness of being, and as such it can also cover over its own event of disclosedness, i.e., not only by remaining with a metaphysics of presence as the approach to the question of being and all events of beings, but by understanding its very event as the function of a mimetic tool that gains its sense of being and direction out of making images of ever-present, unchanging Being as well. However, when Heidegger speaks in the passage above of not telling a story about beings, he is pointing to another aspect of language that leads well beyond its metaphysical interpretation (as a mimetic tool) or function. According to Heidegger, language has the possibility of engaging the disclosedness of being, i.e., in terms of *Being and Time,* of engaging the question of being in its essential temporality and finitude. Heidegger's discourse in *Being and Time* is led by this task. Echoing Aristotle (his *logos apophantikos*), Heidegger introduces his project with a discussion of *logos* and *apophansis.*[32]

At the center of *Being and Time* language appears in its disclosive aspect. The book's project itself occurs as a phenomenology rooted in the disclosive or "apophantic" power of language. In his introductory remarks on phenomenology Heidegger writes,

> When we envisage concretely what we have set forth in our interpretation of "φαινομενον" and "λογος," [*phainomenon* and *logos*] we are struck by an inner relationship between the things meant by these terms. The expression "phenomenology" may be formulated in Greek as λεγειν τα φαινομενα [*legeint a phainomena*] where λεγειν [*legein*] means αποφαινεσθαι [*apophainesthai*]. Thus, "phenomenology" means αποφαινεσθαι τα φαινομενα [*apophainesthai ta phainomena*]—to let that which shows itself be seen from itself in the very way in which it shows itself from itself.[33]

32. To what extent is Heidegger's project driven by Aristotle's relentless project of rendering beings intelligible in their being? To what extent is this force a continuation of a phenomenological focus that aims to make intelligible beings? To what extent is intelligibility the focus of Heidegger's analysis? How close does Heidegger's project come to the metaphysical emphasis on presence in this drive for intelligibility? These are questions that subtend *Being and Time* as well as Heidegger's relation to Aristotle, and that as such must remain at least in the background of this discussion.

33. *BT,* 58, M&R translation (*SZ,* 34).

Phenomenology is composed of two ancient Greek words, *phainomenon* and *logos,* and it takes its direction from the apophantic character of its particular discourse. That the language of phenomenology is apophantic means that it has the possibility of letting beings show themselves out of themselves. This is present in the etymology of "apophantic" and "phenomenon," since both share root meanings. Phenomena and apophantic come from the verb *phao,* "to give light" or "to shine." At the same time, apophantic has a second meaning central to Heidegger's project. It comes also from the verb *phemi,* "to say" or "to make a statement."[34] The phrase *apophainesthai ta phainomena* indicates a possibility of a saying that lets occur the shining forth of events of beings in their disclosedness. The disclosedness of being occurs when the logos in its events lets beings show themselves out of themselves.[35]

Although the passage above recalls Aristotle's terminology (i.e., *logos apophantikos*), Heidegger's discussion points to a way of engagement of events of beings that, unlike Aristotle's thought, does not limit itself to objective and ideal presence. This is indicated by Heidegger's elaboration of the Greek in his last sentence, "to let that which shows itself be seen from itself in the very way in which it shows itself from itself." It is not a matter of making an image, of a mimetic event. The philosophical discourse has the possibility, and hence the task, of engaging beings in their manifestations, i.e., not as already constituted entities (and their unchanging essences) that may be represented. The matter of this logos is the presencing of beings as such. Although crucial, this differentiation between Aristotle and Heidegger does not yet indicate how the latter will engage those issues of alterity and exilic grounds that remain beyond Timaeus' likely story and Aristotle's *logos apophantikos.*

Language and Alterity

In *Being and Time* Heidegger takes up the question of language precisely at the limit of the disclosedness of beings, i.e., in his analysis of the question of being in dasein's (being-t/here's) being-in-the-

34. Liddell and Scott, *A Greek-English Lexicon* (Oxford: Oxford University Press, 1989).

35. This passage recalls Aristotle's *logos apophantikos,* and indicates Heidegger's interpretation of Aristotle's language, an interpretation at variance with the traditional reduction of Aristotle's thought to rational principles of logic and rules of assertive language.

world.[36] In §34 of *Being and Time,* entitled "Dasein and Discourse. Language" ("*Da-sein und Rede. Die Sprache*"), Heidegger discusses "language." Already one finds in the title a telling detail: namely, that dasein is associated with discourse (*Rede*).[37] Then, separated by a period, appears the word "language" (*Die Sprache*). This title makes clear that dasein is to be discussed together with the question of discourse (*Rede*), suggesting as well that language too is to be discussed. But the connection between the first two terms and language is held in suspension by a period that marks their separation, a seeming separation between language, on the one hand, and dasein and discourse, on the other.[38]

If one listens attentively to Heidegger's words in this section, a certain change in the very possibility of understanding the phenomenon of language begins to be heard. The separation previously noted between dasein's discourse and language points in two directions. On the one hand, it is not clear how and in what sense language will be broached as such by moving from discourse to it. On the other hand, when Heidegger brings language back to dasein's discourse he seeks its sense, not in logical, grammatical, or linguistic terms, but in the prerational and concrete experience of dasein's being-in-the-world. Together, these two points indicate that in this section the discussion of the foundations of language in dasein's discourse broaches an understanding of language in an entirely different way from the traditional interpretation of it in terms of objective and ideal presence and as a mimetic and representational tool.

The relationship between discourse and language is quickly introduced in the first passage of the section. Near the end of the first passage Heidegger states that "[*t*]*he existential-ontological foundation of language is discourse.*"[39] This means that in order to understand what he means by language we have first to know more about discourse. Discourse, as the title *Da-sein und Rede* suggests, belongs intrinsically to dasein's being. In the same section Heidegger indicates that "[a]s the exis-

36. The point of adding the "being-t/here" parenthetically is to call attention to the way "in" indicates not something like an entity within which something occurs, but rather dasein as a temporal event.

37. The term "discourse" is used in this section with specific reference to Heidegger's technical term, whereas throughout the book it is used in its more general sense of a specific discussion and language.

38. The period also grammatically announces, commences the topic of language. The period, in calling for a rest, for silence, opens a moment that lets us consider not language as discourse but language as such.

39. *BT,* 150, italics in original (*SZ,* 160–61).

tential constitution of the disclosedness of dasein, discourse is constitutive for the existence of Dasein."[40] By "constitutive" Heidegger means that discourse is the way in which being-in-the-world comes to pass. "The attuned intelligibility of being-in-the-world *is expressed as discourse.*"[41] At this point, a step back is required in order to better grasp this statement.

According to Heidegger's thesis in *Being and Time,* dasein always occurs as being-in-the-world. "Dasein is never 'initially' a sort of a being which is free from being-in, but which at times is in the mood to take up a 'relation' to the world. This taking up of relations to the world is possible only *because,* as being-in-the-world, Dasein is as it is."[42] Dasein occurs always as "something," as an intelligible event, together with other beings, before any conceptual interpretation or assertion. "Intelligibility is also always already articulated before its appropriating interpretation."[43] Being-in-the-world occurs as a kind of prerational ordering. Discourse is the self-articulation of this intelligible event. In this sense, discourse is not the expression (mimesis or representation) of something other than beings in their manifestations, but it is constitutive of the disclosedness of being as enacted in dasein's (being-t/here's) being-in-the-world.

Heidegger's use of "discourse" refers in one sense to the structure that belongs to dasein's disclosedness in its being-in-the-world. Dasein's disclosedness is enacted in discourse and occurs as a structure of call and answer. As Charles Scott points out, "'to call' is helpful in interpreting how a being happens. It is present as something, never as nothing at all, and as something a being occasions an immediate range of possibilities for affirmation, negation, indifference, modification, explanation, appreciation."[44] Being occurs as a call because dasein can only be as being-in-the-world, and this means always being in openness with the disclosedness of other beings that are constitutive of dasein's worldly event.[45] These other occurring beings are not the same as dasein, nor are they determined by dasein; they are in their difference essential to dasein (being-t/here), and, in their being essential and other, they are always calling dasein to its sense of being. Furthermore, these events of beings

40. *BT,* 151 (*SZ,* 161).
41. *BT,* 151, italics in original (*SZ,* 161).
42. *BT,* 53–54, italics in original (*SZ,* 57).
43. *BT,* 150 (*SZ,* 161).
44. Charles Scott, "Heidegger, Madness, and Well-Being," in *Martin Heidegger: Critical Assessments,* ed. Christopher Macann (New York: Routledge, 1992), 281–84, italics in original.
45. "[H]uman existence happens as openness with the immediate presence of beings." Ibid.

say themselves, i.e., in occurring as something, they have always already enacted their being.

At the same time, calling occurs in the openness of an answer. Dasein is not only "there," but is also always a particular being, a someone, and in being this someone, in always having a sense of being, is always in a process of answering the call of being. "Dasein is a being that does not simply occur among other beings. Rather it is ontically distinguished by the fact that in its being this being is concerned about its very being. Thus it is constitutive of the being of Dasein to have, in its very being, a relation of being to this being. [. . .] It is proper to this being that it be disclosed to itself with and through its being. *Understanding of Being is itself a determination of being of Dasein.*"[46] Dasein may refuse, ignore, or turn away from being but it always does so with its sense of being something, always already engaged in the calling and responding that is the discourse of being-in-the-world, a discourse that is always already there as the preconceptual and intelligible condition for the possibility of any assertive judgment or intentional decision.[47] As such, this discourse does not belong to the logic of entities objectively present at hand, nor to their logical necessity, as it occurs rather as their possibility.

Discourse also links dasein's disclosure to the spoken word. In terms of the phenomena the point seems self-evident: dasein always speaks, utters words, signs, symbols. According to Heidegger, the intelligible disclosure of being-in-the-world sets up a "general totality of signification" that occurs in discourse by being put into words. "The totality of significations of intelligibility is *put into words.*"[48] This means that the "word" belongs to the disclosure of dasein. The intelligibility that occurs out of dasein's being-in-the-world is ultimately gathered in words. It is in the sounding out of words that the intelligible disclosedness of being-in-the-world is brought forth, and preserved. Heidegger points out that "[w]ords accrue to significations. But word-things are not provided with signification."[49] Primarily, words grow out of the disclosedness of dasein. And primarily, they are neither things nor informed by meaning from elsewhere, but enact and mark the manifestations of beings in their diversified disclosedness.

46. *BT,* 10 (*SZ,* 12).

47. "The statement as a mode of appropriation of discoveredness and as a way of being-in-the-world is based in discovering, or in the *disclosedness* of dasein." *BT,* 207 (*SZ,* 226).

48. *BT,* 151 (*SZ,* 161).

49. Ibid.

Immediately after his observation regarding words, Heidegger states that "[t]he way in which discourse gets expressed is language."[50] Here Heidegger is speaking of language in a sense that moves outside the traditional understanding of it in terms of the necessity of objective logic. Yes, says Heidegger, language may be interpreted as a body of words in the world, a body that has a being of its own and that may be experienced as one more body of rules in the world. Yes, this language may be broken up into particular significations, word-things. It may be interpreted in terms of grammatical and logical rules.[51] But he then turns the question of language toward the question of the disclosedness of being. As a body of words, language articulates the disclosedness of dasein's being-in-the-world. Language echoes the disclosedness of dasein by being composed of words that gather these expressions into a body of meanings present in a world.[52] Language may be understood not as otherwise than dasein but as an element of dasein's articulate, intelligible disclosedness. In the *Prolegomena,* where Heidegger discusses language at greater length than in *Being and Time,* he points out that "[l]anguage itself has Dasein's kind of being. There is no language in general, understood as some sort of free floating essence in which the various 'individual existences' partake."[53]

In this section, then, Heidegger seeks to understand language out of being's disclosedness in dasein's being-in-the-world. Thus, he moves from discourse as a prerational event to the sounding out of words, to the body of these expressions constituting language. In seeking the foundation of language in dasein Heidegger transfigures the question of language as a question of objective logic and grammar into a question of articulation and mode (dasein's way of being, or being-t/here's being-in-the-world) of the disclosedness of being. In other words, language no longer appears as the logical ordering rule for being, but is now at the service of being.[54] A further look at this transformation shows that in

50. Ibid.

51. *BT,* 155 (*SZ,* 165).

52. In the first page of the "Letter On Humanism"(1946) Heidegger writes, "*Die Sprache ist das Haus des Seins.*" *Wegmarken, GA* 9, 311. One might keep in mind the question of the translation of the sense of "language" that occurs through Heidegger's discussion of it in light of the essential temporality of beings and thought. Thus, this later statement might appear as a question, in the sense of pointing toward the continuity of the thought that began in *Being and Time,* and in the sense of pointing to a sense of language that might be possible only after *Being and Time.*

53. *GA* 20, 373.

54. Cf. *BT,* 69 (*SZ,* 96).

Heidegger's analysis of discourse, dasein, and language, a difference opens between discourse and language, wherein appear possibilities for understanding language that point beyond presence and representation. These possibilities ultimately indicate the alterity of language as figured in Heidegger's project.

According to Heidegger's analysis, language echoes dasein's being-in-the-world. But what is echoed is nothing other than the event of disclosedness in discourse. Heidegger indicates this much when he states that "language itself has Dasein's kind of being." However, this doubling of the disclosedness of being into discourse and language requires further discussion and clarification. First of all, this does not mean that there is a body of linguistic, semantic, and logical rules already present that belongs to an entity at hand called dasein. As is clearly indicated above by Heidegger's words, language is not to be understood in such an objective, conceptual way, nor is dasein a subject or entity among other entities in the world. Language has dasein's kind of being because it arises in the event of dasein's being-in-the-world, and as the echoing of this disclosedness. This means that language arises in and as an elemental part of, the discursive event that is dasein. Furthermore, this discourse is not dasein's but an enactment of the difference that is dasein in its being always in the open with other beings. Only in such prelinguistic difference does dasein come to pass, and only out of that same prelinguistic difference does language come to perform the echoing of disclosedness. Hence as the term "echo" suggests, language occurs not as an imaging of disclosedness but as an elemental aspect of such events.

In light of this analysis some observations toward the alterity of language, and therefore the alterity of the philosophical discourse, can be made. These observations carry this discussion beyond Heidegger's analysis by following some of its implications. Inasmuch as language arises in the call, in prelinguistic openness, language will remain beyond its saying: there will not be a sentence that makes intelligible the originary event of language. At the same time language is not other than the disclosedness of being in its performative or echoing events. Therefore, one can say that language, in echoing, conceals its event. In other words, language will always remain to a certain extent silent concerning its events, and in this sense will remain strange, foreign to its event in its sounding out. To language belongs the presencing and intelligibility of beings in the echoing of being, but this echoing will always occur as an event in alterity, an event beyond its determination and presencing.

It should be noted here that since language is not something outside the disclosedness of being (dasein), this alterity does not simply indicate a failure of language toward being, but as the echoing/enactment of the disclosedness of being it points to the alterity of being in its events of disclosedness. This does not mean that language cannot engage its event, or the disclosedness of being, but it does suggest that such engagement would only occur in light of a certain engagement of the withdrawal in the saying of language. Perhaps, as will become apparent, the most obvious move in this sense is Heidegger's turn to hesitation and silence as operative forces of language (understood in its broadest sense).

Language and the Truth of Being

The alterity of language and its role in the disclosedness of beings in the question of being opens a path to the heart of Heidegger's thought in *Being and Time.* If one follows this trace of alterity, Heidegger's project of apophantic logos reveals itself as a thought enacted in alterity and on exilic grounds, i.e., in light of the impossibility of returning to ever-present, unchanging origins and the impossibility of a representation of the logos in which manifestations of beings come to pass.

In *Being and Time* Heidegger points to this broader understanding of language in terms of discourse by discussing the Greek *logos.*[55] Heidegger recalls Aristotle's famous line that man is *zoon logon echon,*[56] the being who has logos. Then, he explains that this does not mean being an *animal rationale* or having a body of words and grammar at hand. Heidegger interprets Aristotle's phrase to mean that man is "in the mode of discovering the world and Dasein itself."[57] The Greeks did not have a word for language in the sense of a body of words and rules present at hand. They understood logos in the sense of discourse (*Rede*), i.e., as not only measuring, but also gathering and speaking. This means that for the Greeks language occurred as inseparable from being in the world; logos in its fuller sense is being-in-the-world-in-discourse. In other words, the

55. *BT,* 154–55 (*SZ,* 165).

56. On Heidegger's interpretation of Aristotle's statement, see *Platon: Sophistes, GA* 19, 17–19; his discussion of Aristotle's *Nicomachean Ethics* in *GA 19,* 21–64; and *Parmenides, GA* 54, 94–104.

57. *GA* 19, 17–19.

openness of logos echoes the openness of dasein in its being-with (*mitsein*),[58] dwelling (*wohnen*),[59] being along (*sein bei*),[60] and in-being (*Insein*), in being-t/here (dasein) in the open with beings.

According to Heidegger the essence of logos is characterized by the single power of concealment and disclosure. Speaking of the insight he claims to have found in the *Nicomachean Ethics*,[61] Heidegger says, "Aristotle says that the λογος [*logos*] is the kind of being of Dasein which can either uncover or cover up. This *double possibility* is what is distinctive in the truth of the λογος; it is the attitude which can also cover over."[62] What is crucial in this passage, as Heidegger himself emphasizes, is that this double play is the possibility for the truth of a logos in the disclosedness of being. This double possibility indicates that philosophical discourse must keep guard against the covering over of being by the metaphysical and transcendental assertiveness of speeches grounded on objective and ideal presence. A dimension of concealment, of necessary absencing, loss, and withdrawal always accompanies the possibility of disclosedness, as enacted by the logos. It is this withdrawal operative in the truth of any apophantic logos that figures the alterity of Heidegger's project and thought.[63]

In "A Dialogue On Language," Heidegger engages the question of language in its withdrawing dimension.

> JAPANESE: We Japanese do not think it strange if a dialogue leaves undefined what is really intended, or even restores it back to the keeping of the undefinable.
>
> INQUIRER: That is part, I believe, of every dialogue that has turned out well between thinking beings. As if of its own accord, it can take care that that undefinable something not only does

58. *BT* (*SZ*) Part I, Chapter IV, "Being-In-the-World as Being-With and Being a Self: The They."

59. *BT*, 51 (*SZ*, 54).

60. Ibid.

61. See "Phänomenologische Interpretationen zu Aritsoteles: Anzeigen der Hermeneutischen Situation," in *Dilthey-Jahrbuch für Philosophie*, vol. 6 (Göttingen: Vandenhoeck & Ruprecht, 1989). See also *Platon: Sophistes*, *GA* 19, Part I.

62. *BT*, 207, my translation (*SZ*, 226).

63. The withdrawal of being that occurs with the history of metaphysics is not the same as the withdrawal of being operative in all presencing of beings. The latter is left to oblivion by metaphysics, and it figures the truth of being in its presencing/abscencing. For the difficult question of the play of the withdrawal of being in metaphysics and the truth of being as presencing/abscencing, see Heidegger's *Contributions to Philosophy*.

> not slip away, but displays its gathering force ever more luminously in the course of the dialogue. [. . .] Thirst for knowledge and greed for explanations never lead to a thinking inquiry. Curiosity is always the concealed arrogance of a self-consciousness that banks on a self-invented *ratio* and its rationality. The *will* to know does not *will* to abide in hope before what is worthy of thought.[64]

In this passage Heidegger takes issue with the language of metaphysics and presence, and with the absence of a certain "reticence" (*Verschwiegenheit*)[65] in the assertive self-enactment of this metaphysical tradition. The assertive insistence on objective and ideal presence ignores the possibility of engaging beings in their withdrawal. What is lost is the possibility of enacting a certain return to the absencing that is operative in all philosophical discourses and in language in general, a return enacted in reticence and leading back to silence. This possibility of engaging the alterity of events of beings and language is not only missed by the language of presence, but, in its self-perpetuation (for according to the metaphysical understanding of being as ever-present, consciousness can only exist by constantly asserting its presence) this language also covers over the hesitation and silence that belong to language and thought.

This particular passage delineates a powerful characteristic of the logos: to it belongs the possibilities of engaging occurrences of beings in their alterity and of speaking in the awareness of the withdrawing operative in the very events of a philosophical logos or thought. "*Hearing* and *keeping silence* are possibilities belonging to discoursing speech."[66] That listening and silence belong to discourse means that not only the assertive presence of saying, but also the undefinedness of not saying, both belong to the logos as disclosedness. "Another essential possibility of discourse has the same existential foundation, *keeping silent.*"[67] Heidegger also states that "[i]n order to keep silent, Dasein must have something to say, that is, must be in command of an authentic and rich disclosedness of itself."[68] Therefore, "not saying," hesitating, or returning to silence, do

64. Hertz, *On the Way to Language,* 13.
65. *BT,* 154 (*SZ,* 164).
66. *BT,* 151, italics in original (*SZ,* 161).
67. *BT,* 154, italics in original (*SZ,* 164).
68. *BT,* 154 (*SZ,* 165).

not have the sense of nonmeaning. Not saying means keeping silent in the sense of hearing, and keeping silent in light of what must be said. This suggests a speaking with a certain "reticence," which, in its not saying, intimates something of being that is beyond objective and ideal presence and its assertive discourses.

Indeed, for Heidegger, the question of the essences of language and being is held together by a silence that belongs to the occurrences of beings in the truth of their disclosedness. In *Being and Time* Heidegger states that "The being true of the λογος [*logos*] as αποφανσις [*apophansis*] is αληθευειν [*aletheuein*] in the manner of αποφαινεσθαι [*apophainesthai*]—to let beings be seen in their unconcealment (discoveredness), taking them out of their concealment. . . . Thus unconcealment, α–ληθεια [*a-letheia*], belongs to the λογος [*logos*]."[69] Language, for Heidegger, is the possibility of the disclosure of the truth, or unconcealment, of being. The logos belongs to truth when it is a letting-be-seen of beings in their being, or in other words, when the truth of being is enacted/echoed in language. The key indication of the disclosedness of the logos in the presencing/absencing of beings is the Greek term for truth, *aletheia,*[70] which may be broken down into its two parts: *a* and *letheia* (the *alpha* is a probative mark).[71] Thus, it may be transliterated as "unhiddenness." The root of the word is the verb *lanthano,* "to be forgotten." This verb is related to the word *lethe,* which means "forgetting," or "forgetfulness," and in Homer, "a place of oblivion." The Greek term indicates therefore that in belonging to truth (*aletheuein*) logos has the power of "unhiddenness," of bringing out of forgetfulness by letting the occurrences of beings come to presence in their events. But *aletheia* is not only a disclosure in the sense of letting beings come to presence in their events: To the logos of the truth of being belong both the unconcealment of the *alpha* and the hiddenness of its lethic element.[72]

69. *BT,* 202 (*SZ,* 219). Cf. *BT,* 28–30 (*SZ,* 32–34); and "logos und aletheia," in *Plato: Sophistes, GA* 19, 181–88.

70. A full discussion of the terms *aletheia* and *lethe* may be found in Heidegger's *Parmenides, GA* 54. He also worked on this question in the 1930s, in his discussions of *a-letheia,* where the "lethic" element of truth is brought to the question of the truth of being. Cf. *Contributions to Philosophy,* 230–31; GA *65*, 329.

71. Compare with the "er" in *Erschlossenheit,* which here is translated as "disclosedness."

72. For Heidegger's further work on this question, see "Aletheia (Heraklit, Fragment 16)," in *Vorträge und Aufsätze* (Pfüllingen: Neske, 1985), 163–66; and "Aletheia (Heraclitus, Fragment B16)," in *Early Greek Thinking,* tr. David Farrell Krell and Frank A. Capuzzi (New York: Harper and Row, 1975), 114. Cf. *Heraklit, "Der Anfand des Abendländichen Denkens. Heraklit," GA* 55, 127–31.

A certain loss belongs to the *logos apophantikos* in the disclosedness of beings, a forgetting or withdrawing that is operative in events of beings. This is why the echoing of being in language occurs not only as an affirmative announcing but also as an articulate return to silence that recalls the withdrawal operative in the articulate manifestations of beings in their presencing and absencing.[73] The question of the "lethic" (*lethe*), of forgetting, does not refer only to a covering over or leaving to oblivion (*lanthanei, epilanthanomai*).[74] As *a-letheia* indicates, the *lethe* is operative in the disclosedness of events of beings in their presencing. It is because of this lethic or withdrawing aspect of the truth of being that language remains in alterity in its manifestations. It is for this same reason that a philosophical discourse that engages the disclosedness of the occurrences of beings should enact a return to silence or unsaying, thus enacting and echoing the truth (disclosedness) of the occurrences of beings in their withdrawal, *a-letheuein*. In other words, the difficulty of language is not simply that it may reduce being to presence. The further difficulty is that disclosedness (*Erschlossenheit*) will require that being, and philosophical discourse, remain to a certain extent beyond what is said, and so beyond presence and representation. Here Heidegger's project points to its very alterity, since, simply stated, his discourse will have to remain to an extent foreign and beyond its event. Furthermore, the task of the discourse will be a certain return to this alterity. In order to further understand the alterity of Heidegger's project it will be helpful to look at the question of temporality, the central point in Heidegger's understanding of the truth of being in *Being and Time*.

Language and Temporality

How is one to understand the lethic element of the truth of being? The difficulty figured by this question is that of engaging the alterity of being and of philosophical discourse by taking up the withdrawing or absencing operative in the presencing of beings, and to do so without giving the character of "some-thing" (objective or ideal) to the lethic. This should also include a warning against taking this lethic element to be a kind of

73. *BT*, 207 (*SZ*, 226).
74. *BT*, 201–4 (*SZ*, 219–221).

negative and unsayable substance analogous to the negative presence of god in negative theology. The lethic does not refer to any kind of conceptual or transcendental being outside the occurrences of the disclosedness of events of beings, but instead figures the temporality of occurrences of beings in their finite events.[75] In short, for Heidegger, to speak of disclosedness is to speak of manifestation in its full complexity and difficulty. This aspect of disclosedness is broached in *Being and Time* when Heidegger makes his differentiation between historiography (*Historie*) and history (*Geschichtlichkeit*).[76] This differentiation calls attention to fundamental or ontological occurrences of beings as concrete and finite events and to the historiographical engagement of temporality in terms of entities, facts, and the concepts that ground such interpretations. This emphasis on finite events is clearly found in Heidegger's discussion of the "moment" (*Augenblick*), where the truncated temporality of beings (past-present-future) is gathered in the event of dasein's temporal and finite disclosedness or manifestation, i.e., in dasein's concernfull (*Besorge*) being-in-the-world and care (*Sorge*) in the modality of being toward death.[77] Here Heidegger's understanding of the moment does not break down time but takes up the three-fold structure of the moment of dasein in its futurity or being-toward-death. This futurity bears with it a certain withdrawal, insofar as it reveals that presencing is always an event in light of a "not yet"—a projection that is never present and yet is necessary for presencing to occur. The lethic refers to the "not yet" and the loss that are at play in every configuration of beings. What I want to indicate in referring the lethic to the analysis of being in terms of temporality in *Being and Time* is that the lethic is nothing other than the withdrawal operative in all intrinsically temporal configurations of beings, both as loss and as a "not yet" operative in all manifestations.

75. Here I put the question of the lethic in terms of the full project of *Being and Time,* including the third part, where it is Heidegger's aim to think the being of beings out of the disclosedness of being.

76. "Historie aber—genauer Historizität—its [ist: KM] als Seinsart des fragenden Dasein's nur möglich, weil es im Grunde seines Seins durch die Geschichtlichkeit bestimmt ist." *BT,* 17 (*SZ,* 20). This statement is not developed out of the same attunement to the question of being that occurs in the discussions of *Geschichte* in the 1930s in Heidegger's *Contributions to Philosophy, GA* 65; and *Geschichte des Seins, GA* 69.

77. *BT,* 297–304 (*SZ,* 323–31), "Temporality as the Ontological Meaning of Care." In the last section of this chapter I show how this withdrawing element of the disclosedness of being is figured in the project of *Being and Time* with dasein's being-toward-death.

Heidegger's project in *Being and Time,* his apophantic logos, may be understood in terms of the temporality and withdrawing fundamental to events of beings in their manifestations. Such a reading ultimately indicates the exilic character of Heidegger's thought in *Being and Time.* The foregoing discussion of the presencing/absencing operative in events of beings recalls thinking to its temporality, i.e., it recalls thinking to its own passing character or finitude, and to its loss and alterity. As already indicated, the question of language and the disclosedness of being are inseparable in the difficulty of the articulation of the disclosedness of being. This means that philosophical discourse, and particularly Heidegger's apophantic logos, will occur in this double play of disclosedness as both unconcealing and concealing. In light of the lethic character of the truth of being, the logos will only occur as disclosedness in the engagement of the withdrawing or loss operative in events of beings. In the attempt to engage the question of being, thought must occur with a certain attentiveness and reticence in light of the difficulty of the question of disclosedness in the truth of being. The alterity of events of beings and of the logos calls not for a discourse that remains outside the truth (unconcealing and concealing) of being, but rather for a philosophical logos that engages events of beings as it enacts its loss in the events of disclosedness, i.e., by writing out of the awareness of its own lethic dimension. In the engagement of the alterity of the disclosedness of beings, the logos must remain open to its loss. But this necessary task of philosophical discourse leads it outside its already operative conceptual determinations, since it will not be in terms of these determinations already present that the discourse will be engaged in its withdrawing or concealing event. As Heidegger indicates, it is in remaining open in a certain indeterminacy, and by returning to silence, that the logos begins to engage thought's manifestations in their truth (unconcealment and concealment), and hence in their withdrawing or alterity. It is this necessary openness that will call philosophical logos beyond its determination and will recall its alterity.

According to Heidegger, the reticence of the logos that seeks its disclosedness belongs to the most powerful form of speaking, one that may be found in poetry, where dasein's very possibility is articulated in its temporality. "The discoveredness of Dasein, in particular the disposition of Dasein, can be made manifest by means of words in such a way that certain new possibilities of Dasein's being are set free. Thus discourse, especially *poetry,* can even bring about the release of new possibilities of the being of Dasein. In this way, discourse proves itself positively *as a mode of maturation, a mode of temporalization* of Dasein itself."[78] As indicated above, according to Heidegger's analysis, discourse is the articulate enactment of dasein's disclosedness in its being in the open with beings. This event and its echoing aspect (language) occur as presencing and absencing. The engagement of this structure of disclosure in its alterity opens possibilities for beings leading well beyond already operative configurations of the senses of beings. Poetry, if understood in terms of this disclosedness and not merely as a mimetic art, bears such possibilities. The passage above not only indicates that language opens up a possibility for an unchanging dasein, and, therefore, for the occurrences of beings; it also marks Heidegger's recognition of language as a matter of "new possibilities of the being of Dasein." Discourse and language, both enacting dasein's disclosedness and understood in their temporality, open dasein and all events of beings to figurations of being that are yet to come. Here, "yet to come" does not imply something waiting on the horizon, but highlights the temporality operative in occurrences of beings and in their conceptual determinations. As such, the phrase indicates a certain absence in all determinations, an aspect of the withdrawal operative in all disclosedness—a withdrawal, in this case, engaged in its futurity. "Not yet" (*noch nicht*) indicates something that lets beings appear in a projection toward their future, a projection that must remain a nonpresence although operative in the presencing of events of beings and of conceptual determinations. This open horizon for events of beings occurs, for Heidegger, in the work of certain poets, as "a mode of temporalizing of Dasein." Here, the possibilities of events of beings, understood in terms of temporality or dasein's being-toward-death, appear neither in terms of ever-present, unchanging origins and principles, nor in terms of a teleology that establishes a set of possible and necessary determinations for the senses of events of beings. Discourse, or dasein's disclosedness, understood in its temporalizing event, opens beings to their events in a leap that goes beyond already operative configurations and senses of being. This disclosedness beyond presence and representation is what dasein as discourse opens, what poetry enacts, and this is the task of Heidegger's apophantic logos in his engagement of the disclosedness of beings in their truth or operative alterity.

78. *Prolegomena, GA* 20, §28, italics in original.

Exilic Spaces of Thought

Considering the alterity of the disclosedness of being and the way philosophical discourse must engage this event, the task of an apophantic logos appears ultimately as a task that occurs in and enacts a certain exilic event. As I indicated in the introduction to this work, the term "exilic" figures both the impossibility of claiming any sense of identity or sense of being in terms of unchanging, ever-present origins, and a sense of being that remains always open toward the configurations of senses of being not yet given determination.

Heidegger's understanding of the disclosedness of being and his project of an apophantic logos that engages this disclosedness in its alterity lead to such an exilic thought or space of engagement of the occurrences of beings. First of all, the alterity of language always puts the philosophical logos beyond a claim or return to its origins. This is because of the loss or withdrawal operative in all language. At the same time, this withdrawal also makes a claim to any ever-present or unchanging origin impossible. In short, in light of the alterity of language, philosophical discourse will enact a loss of a claim to ever-present, unchanging origins. Second, the futurity of discourse in the opening of possibilities for beings not only repeats the alterity of philosophical logos in its determinations with respect to the possibilities it opens (the "not yet" that cannot arrive since it must remain a future operative in presencing), but it indicates the way that philosophical discourse remains open to unexpected configurations of beings, since all determinations have the character of open events, and hence of possibilities to come. As such, Heidegger's apophantic logos offers a certain openness to determinations of beings, an openness that can not preclude a teleological orientation or an identity precluded by objective and ideal presence alone.

Conclusion

Heidegger's project in *Being and Time* indicates a certain exilic aspect of his thought. At the same time, spatiality and alterity appear as operative aspects of the question of being. This certainly echoes the difficulties figured by Timaeus' *chora* as well as by Aristotle's rethinking of this figure as place (*topos*). Heidegger's project also echoes Aristotle's thought by taking up the term apophantic logos. However, when Heidegger takes these terms as points of departure for his project he also appropriates them and takes them well beyond Plato's and Aristotle's philosophical discourses. Unlike Timaeus' likely story and Aristotle's *logos apophantikos,* Heidegger's project of an apophantic logos aims to rekindle the question of being and, at the same time, leads toward the engagement of the alterity and exilic character of events of beings and thought. For Heidegger language is not a mimetic tool, its power is not its ability to represent or make intelligible the Being behind occurrences of beings. The ground of language, and hence of Heidegger's thought, is dasein's temporality and its not objective and ideal presence. Its force is in the enactment of the disclosedness of beings; a force found by an understanding of language that commits it, and with it, Heidegger's project, to the task of engaging the absencing or withdrawing that is operative in events of beings, and thereby to the difficulty of the alterity of these events in their disclosedness. Heidegger sets out with a conception of language that calls for the engagement of the alterity of beings and thought from the start. The discussion of Heidegger's understanding of dasein, discourse, and language as disclosedness indicates that his project arises in light of alterity and is invested with the alterity of its event. Unlike the ancients, Heidegger does not remain with objective and ideal presence alone in order to engage the occurrences of beings in their disclosedness. Rather, in taking up the question of being in its temporality, he begins to engage the difficulty of alterity that appears in Timaeus' likely story and Aristotle's forceful account of spatiality in the *Physics.* In light of Heidegger's powerful project a number of questions arise. Does Heidegger's thought in *Being and Time* engage the disclosedness of beings in their alterity? To what extent does his language take up the alterity of its event? Does Heidegger's thought engage the exilic character of his thought as figured by his understanding of disclosedness as an event that bears both loss and a certain transformative force in its futurity? The next four chapters take up these questions by focusing on the way the figure of spatiality is operative at crucial moments in the thinking of *Being and Time.*

PART TWO

Scherzi

The first section introduced a series of themes intrinsic to the question of being, to the occurrences or events of beings, and to the manifestations of thought. Among them were the following: the issue of spatiality; this issue as figure of the alterity and exilic character of events of beings and of thought; the withdrawal and concealing operative in the disclosedness of events of beings and thought; and finally, interruption, silence, and hesitation as essential elements for the engagement of manifestations of beings and thought in light of their alterity, their exilic grounds, and the withdrawal or truth (revealing and concealing) essential to their presencing. This second section takes up these themes at crucial moments in the thinking of *Being and Time.* The various moments in Heidegger's thought are not only engaged in terms of what they say but also in their performative aspects.

Scherzo (*Scherzi* is its plural form). This Italian musical term refers to a playful passage that explores and opens themes in their varying and colorful resonance, while remaining ephemeral and light.

3

Interruptions

The Twisting Free of Spatiality

Introduction

The preceding chapters introduced the difficulty of thinking the occurrences or events of beings in their alterity and on exilic grounds, and suggested that, unlike Timaeus' likely story or Aristotle's *logos apophantikos,* Heidegger's understanding of language and thought in *Being and Time* opens a new possibility for engaging and thinking out of the alterity of thought. As indicated in Chapter 2, this opening requires the interruption of the metaphysical and transcendental interpretation of occurrences of beings in terms of objective and ideal presence and unchanging origins. Heidegger's thought also requires a break in the assertive language that sustains and is founded by such interpretations of the occurrences of beings. One is faced here with an almost impossible task: to suspend the engagement with the world in terms of entities, to interrupt the interpretation of beings as substances, and to break with the understanding of humans on the basis of a self-consciousness determined by self-evident reason. The engagement of beings in their alterity will depend on these interruptions. Regarding this task, the obvious questions are: How do these interruptions occur in *Being and Time?* How does Heidegger's thought break with the tradition? What path is at least intimated in such a rupture?

This chapter takes up these questions by discussing Heidegger's critique of traditional ontology through his direct critique of Descartes' "world ontology" in Part I, Chapter 3 of *Being and Time.*[1] According to Heidegger, Descartes' understanding of beings as *extensio* is a prime example of the interpretation of the events of beings in terms of objective and ideal presence (being as substance). Here spatiality takes a central role in the critique of traditional ontology presented in *Being and Time,* a critique that not only interrupts the metaphysical and transcendental interpretation of events of beings, but that serves as "a negative introduction" to the discussion of the spatiality of being in the paragraphs immediately after the critique as well.[2] In short, with his critique of Descartes, Heidegger leaves the tradition by interrupting it and indicating a path for a thought yet to come. In this moment of passage spatiality twists free from traditional metaphysics and transcendental philosophy, and at the same time, the alterity and exilic grounds of Heidegger's thought begin to be felt.[3] I shall follow this double path of separation and intimation first by giving a brief sketch of Descartes' ontology as background to Heidegger's critique and then by taking up Heidegger's critique in *Being and Time.*

1. Throughout this chapter I have used quotation marks to differentiate the traditional conception of the "world" as "nature" and "spirit" from the world, which is Heidegger's formal indication of the occurrences of beings as engaged out of dasein's being-in-the-world. Heidegger identifies world as a formal element of being-in-the-world in his introductory analysis of dasein's ontological structure. *BT,* 53 (*SZ,* 57).

2. *BT,* 83–94 (*SZ,* 89–101).

3. "Twisting free" is David Farrell Krell's English rendering of Heidegger's word *Herausdrehung.* Heidegger uses this term to indicate the motion of Nietzsche's thought. See Heidegger, *Nietzsche: "Der Wille zur Macht als Kunst," GA* 43, 251. For a further discussion of these two terms, see also John Sallis, "Twisting Free—Being to an Extent Sensible," in *Echoes: After Heidegger* (Indianapolis: Indiana University Press, 1990).

Descartes' "World"

"World"

On 31 January 1642 Descartes writes to Constantjin Huygens, "My *World* would be out already were it not that first of all I want to teach it to speak Latin. I shall call it the *Summa Philosophiae,* to help it gain a better reception among the Schoolmen, who are now persecuting it and trying to smother it at birth."[4] When Descartes speaks of the *World* he refers to his suppressed universal treatise entitled *Le Monde,* a treatise meant to replace the traditional philosophical and scientific texts based on Aristotle's philosophy.[5] But what is in play here is not a particular book. Rather, as the title of Descartes' book indicates, it is the "world" that is at stake.[6] The *Summa Philosophiae,* mentioned in the letter, is first published in 1644, in Latin, under the title *Principia Philosophiae.* Subsequently, a French version titled *Principles of Philosophy* (*Principes*) appears in 1647. In the preface to the French edition Descartes writes that "the word philosophy means the study of wisdom, and by wisdom is meant not only prudence in our everyday affairs but also a perfect knowledge of all things that mankind is capable of knowing."[7] The book is meant to offer universal knowledge, knowledge of all events of beings. Descartes goes on to state of philosophy, "it encompasses everything the human mind is capable of knowing. Thus we should consider that it is this philosophy alone that distinguishes us from the most savage and barbarous people, and that a nation's civilization and refinement depends on the superiority of the philosophy which is practiced there."[8] Descartes' project is all encompassing. His aim is to present a new and unified understanding of

4. This term comes from the translator's preface to *Principles of Philosophy,* in *The Philosophical Writings of Descartes,* vol. 1, tr. Cottingham, Stoothoff, and Murdoch (Cambridge: Cambridge University Press, 1988), 177.

5. Ibid.

6. The aim of this discussion is not to give a complete interpretation of Descartes' thought, but to introduce it as a way of highlighting Heidegger's critique of Descartes and what it accomplishes.

7. *Principles,* preface (IX: b, 2 [179]). Throughout this chapter I will refer to Descartes' work by presenting the name of the work, the section, the original page number as it appears in the standard twelve-volume edition of Descartes' works (*Oeuvres de Descartes,* rev. ed., ed. C. Adam and P. Tannery [Paris: Vrin, 1964–76]), and the page number in brackets of the English translation (*The Philosophical Writings of Descartes,* ed. John Cottingham et al, vol. I [Cambridge: Cambridge University Press, 1985]).

8. *Principles,* (IX: b, 3 [180]).

all beings; his thought and method will define and determine conceptually the "world." But in what manner is such a universal task to be accomplished?

True philosophy for Descartes begins in securing a way of thinking that will guarantee true and distinct knowledge of all beings.[9] Then, the various parts of philosophy may be known. "Thus, the whole of philosophy is like a tree. The roots are metaphysics, the trunk is physics, and the branches emerging from the trunk are all the other sciences." Descartes' *Principia* is structured according to this schema. The first part of the treatise concerns "'the principles of knowledge,' thus what may be called 'first philosophy' or 'metaphysics,'" i.e., "metaphysics" as the principles of knowledge, the explanation of the attributes of God, the nonmaterial nature of the soul, and the notions that the thinking subject already has within. The other three parts contain what concerns "physics," i.e., the true principles of material things and the composition of the universe, the earth, and bodies, including the particular animals, plants, and man.[10]

In Part I of his *Principia* Descartes identifies God as the only self-sufficient substance, and then divides all created beings into two substances, namely "mind" and "matter." "But I recognize only two ultimate classes of things: first, intellectual or thinking things, i.e., those which pertain to mind or thinking substance; and, secondly, material things, i.e., those which pertain to extended substance or body."[11] The principle attribute of mind is "thought," and the principle attribute of corporeality is "extension." "To each substance there belongs one principle attribute; in the case of mind, this is thought, and in the case of body it is extension."[12] Thought and extension can be recognized as respectively constituting the nature of mind and body. Thus, Descartes' "world" is ontologically divided between "corporeal substance" and "thinking substance."

The whole of Descartes' ontology of the world as presented in Part I of his *Principia* is based on his conception of "substance." "Substance" is primarily defined in terms of the ever-present, unchanging being of God. For Descartes, "no signification of this name 'substance' which would be common to God and his creatures can be distinctly understood."[13] This

9. *Rules,* I (X: 360–61 [9–10]).
10. *Principles,* (IX: b, 14 [186]).
11. *Principles,* I.48 (VIII: a, 23 [208]).
12. *Principles,* I.53 (VIII: a, 25 [210]).
13. "[N]ulla eius "substantiae" nominis significatio potest distincte intelligi, quae Deo et creaturis sit communis." *Principles,* I.52 (VIII: a, 24 [210]).

means that the being of substance in this primary sense cannot be known through beings, since they are created by and dependent on God. "Hence the term 'substance' does not apply univocally, as they say in the schools, to God and to other things; that is, there is no distinctly intelligible meaning of the term which is common to God and his creatures."[14]

But for Descartes substance can be known in a secondary sense, as the "principle attribute" of all beings, as "thought" and "extension." The key term here is "attribute." It is through the "attributes" present in things, in created beings, that substance can be known. What is knowable then is not the being of substance as such (God), but its attributes as they appear through the knowledge of beings. Descartes writes in §52 that "the term 'substance' applies univocally to mind and body."[15] He then explains how a substance is knowable through its attributes. "We can, however, easily come to know a substance by one of its attributes, in virtue of the common notion that nothing possesses no attributes, that is, no properties or qualities. Thus if we perceive the presence of some attribute, we can infer that there must also be present an existing thing or substance to which it may be attributed."[16] Substance is knowable as an "attribute," and this means as a nonsensible "existing thing or being."

For Descartes these attributes are never outside of the mind, but belong to the mind. He writes in §48, "[p]erception, volition and all other modes both of perceiving and of willing are referred to thinking substance."[17] Attributes are known through the created phenomena, but they do not belong to the phenomena; rather the phenomena are composed by the attributes that are present to the mind, which make possible the phenomena. In this sense attributes are intuitions of the thinking substance, where sense perception belongs neither to mind nor body alone but to a synthesis of the two. "But we also experience within ourselves certain other things which must not be referred either to the mind alone or to the body alone . . . and finally, [we experience] all the sensations, such as those of pain, pleasure, light, colors, sounds, smells tastes."[18] The constant presence of the attributes of substances in the mind calls for a specific kind of knowledge. Descartes writes in his *Rules for the Direction of the Mind* that "[w]e should attend only to those objects of which

14. Ibid.
15. *Principles,* I.52 (VIII: a, 24–25 [210]).
16. Ibid.
17. *Principles,* I.48 (VIII: a, 23 [208]).
18. *Principles,* I.48 (VIII: a, 23 [209]).

our minds seem to be capable of having certain and indubitable cognition."[19] The kind of knowledge fitting to the idea of substances is found in mathematical knowledge, "arithmetic and geometry . . . they alone are concerned with the object so pure and simple that they make no assumption that experience might render uncertain."[20] But this does not mean that mathematics and arithmetic are the only sciences that may arrive at knowledge of the substance of beings. Descartes writes that "in seeking the right path of truth we ought to concern ourselves only with objects which admit of as much certainty as the demonstrations of arithmetic and geometry."[21] It is not that Descartes' interest in mathematics determines his idea of substance, but substance in its purity requires a focus only on what is unchanging, ever-present, and certain in all beings, and this is achieved by a thinking akin to mathematical knowledge.

If philosophy is a tree, its roots are in the mind's knowledge of substances through their attributes, and these given attributes can only be brought to light through a thinking that can provide demonstrations as certain as those in algebra and geometry. All knowledge of nature's principles, of particular beings, and of human beings begins in this *mathesis universalis.*[22] Descartes' project has as its task to ultimately articulate all events of beings in terms of presence and through a logical, assertive discourse capable of representing and measuring them. However, in spite of this positivistic unifying ontology, when one looks at Descartes' concept of extension (*extensio*) one finds two different ways of understanding spatiality, both held together by the concept of substance.

World-Space

For Descartes, spatiality has three inseparable names: *extensio,* "space," and "place." *Extensio* is the attribute of all corporeal substance. "Extension in length, breadth, and depth constitutes the nature of corporeal substance; and thought constitutes the nature of thinking substance. Everything else which can be attributed to body presupposes extension, and is merely a mode of an extended thing."[23] At the same time, according to this last statement, "space" is nothing other than this corporeal

19. *Rules,* II (X: 362 [10]).
20. Ibid.
21. *Rules,* II (X: 366 [12]).
22. *Rules,* IV (X: 378 [19]).
23. *Principles,* I.53 (VIII: a, 25 [210]).

substance, i.e., nothing other than *extensio.* Descartes writes, "the nature of a body is exactly the same as that constituting the nature of space."[24] Finally, the term "place" means generally[25] nothing other than "the body which is said to be in a place."[26] Thus, ultimately the three terms refer to one corporeal substance, namely *extensio.*

But at the same time, these terms do indicate a difference. *Extensio* is understood as an attribute of a substance, where "space" and "place" refer to material things. Descartes discusses *extensio* in the first part of his treatise, as a part of "The Principles of Knowledge," while "space" and "place" are discussed in the second part of the treatise, as part of "The Principles of Material Things." This second section is a portion of Descartes' discussion of "all that is most general in physics, namely the first laws or principles of nature."[27] "Space" and "place" are laws of "nature." Spatiality appears as a double: as the material "space of nature," and as the intuited "attribute of the substance of nature." In other words, one finds here a double concept of spatiality presented in terms of the metaphysical dualism of ideal and objective presence.

This split shows a break within Descartes' idea of substance. For Descartes, substance reveals itself to consciousness as the intuited attribute of all things, and at the same time reveals itself as the given attribute of the phenomena of extension, a phenomena substantially other than the thinking substance. One cannot say, as is the case for Kant, for example, that the condition for the possibility of the experience of the "space of nature" is the spatial intuition of consciousness only. Descartes' understanding of spatiality as intuited substance and material substance shows an ontological ambiguity in his concept of substance. How does one understand the way in which the word "substance" is used univocally for both things and their substance?

24. *Principles,* II.11 (VIII: a, 46 [227]).

25. In *The Fate of Place* Casey points out that Descartes' unified concept *extensio* bifurcates in his discussion of "place." For Descartes place has two senses: It may mean "internal" or "external" (*Rules,* I: 10,13, and 15). The first is equated with the substance of things; it is *extensio.* One refers to "place" in this sense when speaking of changing the place of a thing. The latter sense refers only to the spatial relation between bodies: It is not substantial but relational. One uses this sense when speaking of putting something in the place of something else. Casey's point is that ultimately place is not simply equated with extension, as Descartes claims, despite the dualism that appears in Descartes' discussion of place.

26. *Principles,* II.13 (VIII: a, 47 [228]).

27. *Principles,* Preface (IX: b, 16 [187]).

What holds together the "space of nature" and *extensio,* in spite of this break, is the concept of substance as such. For Descartes, spatiality is understood as a natural phenomena grounded on substantiality. It is by understanding *extensio* as the ever-present substance that grounds all spatiality that the split between the "space of nature" and *extensio* is bridged. This does not mend the irreparable fissure between thinking substance and extension, but, as Heidegger will indicate in his critique, it does obscure or cover over the question of being by focusing on being as a substance immediately given to mind and reason.

Heidegger's Critique of Descartes' Ontology of the "World"

Brief Sketch

In Part I, Chapter 3 of *Being and Time* Heidegger gives his an analysis of "The 'Worldliness' of the World." This is an analysis of the world, one of the formal elements of dasein's being-in-the-world, and therefore an essential part of the analysis of the disclosedness of beings in light of dasein's temporality.[28] The analysis is tripartite: the analysis of dasein's environment and world; a contrasting analysis of Descartes' ontology of the "world"; and, finally, a positive account of the spatiality of the environment, the world, and dasein's own spatiality.[29] The middle section of the chapter, "A Contrast Between our Analysis of Worldliness and Descartes' Interpretation of the World," is well known as a critique of Descartes' ontology, and through it, of the tradition that follows it, including Husserl's phenomenology implicitly. At the same time, according to Heidegger, the critique has another function as a contrast to his ontological understanding of the world. As such, the discussion also may serve as a "negative support for the positive explication of the spatiality of the surrounding world [*Umwelt*] and of Dasein itself."[30] Thus, the critique is meant to open the way for another mode of thought, i.e., for

28. The break up of being-in-the-world into formal components for the sake of analysis occurs in §12 of *Being and Time.*

29. *BT,* 62 (*SZ,* 66).

30. *BT,* 83, my translation (*SZ,* 89); *BT,* 78–79, M&R translation (*SZ,* 53–54); *BT,* 207–8, my translation (*SZ,* 226).

Heidegger's own discussion of the spatiality of events of beings. How does this analysis work as a critique? And how does it work as an introduction to a thought other than the tradition's?

Heidegger's discussion of Descartes' "ontology of the world" is a critique of the concept of *substantia,* a concept that grounds the interpretation of the phenomena of the world in terms of objective and ideal presence. As noted, Descartes' ontology presents two substantial kinds of being: material substance (*res corporea*), and thinking substance (*res cogitans*). This duality is developed in Descartes' thought, and thereafter referred to as "nature" and "spirit." Heidegger's critique focuses on Descartes' claim that the being of all beings and the world, gathered under the concept of "nature" and its "principles and laws," is *extensio.* In other words, the aim of Heidegger's critique is to overcome the interpretation of the phenomena of the world as "nature," and to engage the occurrences of beings out of their events, i.e., in their temporality or finitude. In his introduction to the chapter, Heidegger points out that "[n]ature is a limiting case of the being of possible entities within-the-world," a limiting case of the phenomena of the world.[31] "Nature" here refers to the entities present at hand and to the concepts grounding such a way of interpreting events of beings. Later Heidegger elaborates his critique of "nature" as such, and points out that this concept is a narrow way of interpreting the phenomena of the world, which ultimately misses the world altogether. Heidegger states that "Descartes narrowed down the question of the world to that of the thingliness of nature [*Naturdinglichkeit*]."[32] Referring to this interpretation, he remarks at the beginning of the same section that "his [Descartes'] interpretation and the foundations on which it is based have led him to pass over both the phenomena of the world and the being of those entities within-the-world"[33] This chapter seeks to uncover the being of the world, not in terms of objective and ideal presence, and not by referring to things and their unchanging concepts. Instead, Heidegger's aim is to go directly to the phenomena of the world,[34] to the event of the worldliness of the world.

31. *BT,* 94, M&R translation (*SZ,* 65).

32. *BT,* 92, italics in original (*SZ,* 100).

33. *BT,* 128, M&R translation (*SZ,* 95).

34. As already mentioned in Chapter 3, n. 5, one should be careful to distinguish between Heidegger's understanding of "phenomena" in terms of *phaenomena*—beings showing themselves—and Husserl's understanding of "phenomena" as the objects of intentional acts of consciousness. Heidegger is referring to that sense of being that precedes consciousness. On this point, see Heidegger's *Die Grundprobleme der Phänomenologie, GA* 24, §12, 229–42.

Heidegger writes, "[i]f we pose the question of 'world,' *which* world is meant? Neither this nor that world, but rather *the worldliness of the world in general.*"[35]

But the general aim of Chapter 3 will not be accomplished simply by giving a positive account of worldliness. Heidegger has already done this in §A of this chapter.[36] Because the phenomena have been covered over by their narrow interpretation in terms of the principles and laws of "nature" and in terms of objective and ideal presence, part of the uncovering of the phenomena will require the deconstruction of such a traditional interpretation.[37] Heidegger accomplishes this overturning of the "natural" interpretation of the phenomena of the world in his critique of the concept of substance in the middle section of the chapter. Thus, his critique has two immediately evident steps. The first is his exposure of the insufficient grounds for Descartes' conception of "substance," i.e., his critique of the idea that leads Descartes to assert that the being of all beings is their *extensio.* According to Heidegger, Descartes never asks the question of the being of substance. Indeed, he argues in the middle section of his critique that Descartes never so much as raises the question of the being of substance, and of its "substantiality," that he avoids it. Heidegger concludes that ultimately Descartes understands the question as inaccessible to the human intellect, and therefore dismisses it on principle.[38]

The second step occurs in Heidegger's "Hermeneutical Discussion of Descartes' Ontology."[39] In this discussion Heidegger shows that, as a result of the way Descartes interprets the phenomena of the world, he never asks the question of its being. Instead, as Heidegger observes, Descartes prescribes to the occurrences of beings a "true" being founded on his ungrounded idea of substance: "but rather [Descartes] prescribes

35. *BT,* 60 (*SZ,* 64).

36. *BT,* 13–17 (*SZ,* 15–18).

37. I am translating Heidegger's word *Destruktion* as "deconstruction" in order to point to the dual motion of the thought in a simultaneous taking down and building up. This word is meant to point to the bringing forth of the "positive possibilities" of the tradition, and is far from having "the negative sense of shaking off the ontological tradition" *BT,* 20 (*SZ,* 23). The other alternative is to translate it as "destruction," but in this English term there is no connotation beyond the negative image of obliterating and leaving behind. Thus, the transliteration of *Destruktion* into its English cognate "destruction" completely covers over this fundamental aspect of Heidegger's thought.

38. "Descartes not only completely evades the ontological question of substantiality, he emphasizes explicitly that substance as such—that is, its substantiality, is in and for itself inaccessible from the very beginning." *BT,* 87 (*SZ,* 94).

39. *BT,* 88–94 (*SZ,* 95–101).

to the world, so to speak, its 'true' being on the basis of an idea of being (being = constant objective presence) the source of which has not been revealed and the justification of which has not been demonstrated."[40] This serves to complete Heidegger's critique. Not only is Descartes' ontology unfounded, but it also misses the phenomena of the "world" altogether. In this last part of his critique Heidegger does more than show that the phenomena have been "passed over." That the phenomena of the world have been passed over means that they have never been reached. By pointing this out, Heidegger directs thought back to the phenomena of the world. The question of the being of the world must be asked out of the phenomena that have been passed over.

One gets a glimpse of the force of Heidegger's critique by looking at how he takes the question back to the phenomena of world and the occurrences of beings in the last section of his critique, in his "hermeneutical discussion." The term "hermeneutical" traditionally refers to interpreting a text for its message. The word comes to us from the name of the messenger god of the ancient Greeks, Hermes. Hermeneutics initially involved the interpretation of religious texts, the search for the meanings of the sacred word, and has come to refer to the involvement of the reader in the reading. Thus, interpreter and interpreted are gathered in a historical interpretation out of which meaning arises.[41] Heidegger's sense of hermeneutics goes further and refers to the very event that is the possibility of any interpretation at all. In *Being and Time* Heidegger writes:

> The *logos* of the phenomenology of Dasein has the character of *hermeneuein*, through which the authentic meaning of being, and also those basic structures of being which Dasein itself possesses, are *made known* to Dasein's understanding of being. . . . [T]his hermeneutics also becomes a "hermeneutic" in the sense of working out the conditions on which the possibility of any ontological

40. *BT,* 89 (*SZ,* 96).

41. Interestingly, one finds at the beginnings of the development of hermeneutics the differentiation between *Geschichte* and *Historie.* This differentiation does not begin with Heidegger, but is first made by the German theologian Martin Kähler. It is then systematically sustained in scriptural hermeneutics from Wilhelm Dilthey to Rudolf Bultmann. See *Historisches Wörterbuch der Philosophie,* vol. 3, ed. J. Ritter (Darmstadt: Wissenschaftliche Buchgesellschaft, 1974): 398–99. In relation to Dilthey, see Chapter 77 of *Being and Time.* The differentiation between *Geschichte* and *Historie* becomes the central focus in Heidegger's later hermeneutical thought in the *Contributions to Philosophy,* in which Heidegger refers to his thought as a *seinsgeschichtliches Denken* (*GA* 65, 3).

> investigation depends. . . . And finally, to the extent that Dasein, as an entity with the possibility of existence, has ontological priority over every other entity, "hermeneutic," as an interpretation of Dasein's being, has the third and specific sense of an analytic of the existentiality of existence; and this is the sense that is philosophically primary.[42]

Being and Time is a hermeneutics of hermeneutics in that it is an analysis of the interpreting that is the possibility of any interpretation at all. Or, in other words, the interpretation of dasein's being-in-the-world is the hermeneutics of the hermeneutical event that is the grounding of any hermeneutics or interpretation.[43] But what is the significance of this move to the hermeneutic event as such, or, in the specific case of *Being and Time,* to dasein's being-in-the-world?

At the end of his "hermeneutical discussion" Heidegger writes that Descartes' interpretation is ultimately made visible only if understood in light of dasein's being.

> We have already intimated (section 14) that passing over the world and those beings initially encountered is not a matter of chance, not an oversight which we could simply make up for, but rather is grounded in the essential kind of being of Dasein itself. When our analytic of Dasein has made the most important basic structures of Dasein transparent in the scope of this problematic . . . only then can the critique of the Cartesian ontology of the world, basically still customary today, claim its philosophical justification.[44]

The previous passage and this one take the question of the being of the world back to the event of dasein's being-in-the-world. With them emerges an indication of a thought beyond metaphysics and transcen-

42. *BT,* 61–62, M&R translation, italics in original (*SZ,* 37–38).

43. "In *Being and Time,* the term 'hermeneutics' is used in a still broader sense, 'broader' here meaning, however, not the mere extension of the same meaning over a still larger area of application [sic]. 'Broader' is to say: in keeping with that vastness which springs from originary being. In *Being and Time,* hermeneutics means . . . the attempt first of all to define the nature of interpretation on hermeneutic grounds [das Wesen der Auslegung allererst aus dem Hermeneutischen zu bestimmen]." "Aus einem Gespräch von der Sprache," in *Unterwegs zur Sprache* (Pfüllingen: Neske, 1986), 96–98. See also "A Dialogue on Language," in *On the Way to Language,* tr. Peter D. Hertz (San Francisco: Harper, 1971), 9–11.

44. *BT,* 93 (*SZ,* 100).

dental philosophy. The last lines of the passage immediately above move precisely in the direction of the interpretation of the phenomena of the world through the engagements of dasein's being-in-the-world. This move figures the question of the ground for any intuiting consciousness. As shown above, the concept of "substantiality" is the ground of Descartes' ontology, and of his division of being into *res cogitans* and *res extensa.* This means that when Heidegger turns back to the phenomena of the world or the occurrences of beings and to the interpretative dimension of the origin of these events, he is ultimately attempting to uncover the originary event of the *cogito sum* as well as of the concept of "nature"; that is, he is asking the question that transcendental philosophy takes as a given when this tradition understands consciousness as self-evident.[45] This is why he introduces his critique of Descartes' ontology by writing that "[t]he considerations which follow will not have been grounded in full detail until the *cogito sum* has been phenomenologically deconstructed."[46] The issue is precisely the uprooting of consciousness and intuitional knowledge as the ground for any interpretation of the phenomena. This uprooting is a necessary aspect behind Heidegger's statement that one must go back to dasein's being-in-the-world in order to philosophically grasp the concept of substantiality that underlies the ideas of "consciousness" and "nature."[47]

In returning to the phenomena of the hermeneutical event as the grounding of Descartes' conceptuality, Heidegger's critique moves beyond Descartes' ontology. Events or occurrences of beings refer neither to substance, nor to "consciousness," nor to "nature." Rather, they occur as phenomena out of being-in-the-world in its disclosedness. In pointing out in this last section that Descartes' ontology passes over the phenomena, Heidegger also points to the phenomenon that remains to be thought, which is the disclosedness of the occurrences of beings in their events or temporality and finitude. One may say that Heidegger repeats Husserl's insight concerning the intuitional sense of thinking, and that, furthermore, this repetition characteristically reappropriates the insight by taking it further, outside of pure consciousness, and back to its originary grounding events or the world—i.e., to the phenomena of the occurrences of beings and thought in their temporal events and finitude.

45. *Die Grundprobleme der Phänomenologie,* Part I, *GA* 24.

46. *BT,* 123, M&R translation, italics in original (*SZ,* 89).

47. This is one way to indicate at least the way in which Heidegger's critique of Descartes' ontology is simultaneously a critique of Husserl's phenomenology.

Briefly, to recapitulate, the critique of Descartes' ontology of the "world" occurs as a transformative hermeneutics that is a part of the overcoming of transcendental philosophy, and more specifically, as the deconstruction of the concept of *substantia* that grounds the interpretation of the phenomena of the world as "nature" and "spirit." This deconstruction interrupts the interpretation of events of beings in terms of objective and ideal presence, and points toward the phenomena of the world and hermeneutics as the originary events in which all these concepts arise. Heidegger's critique not only serves as an analysis and deconstruction of Descartes' ontology; it also points out the need for an engagement of events of beings in their temporality and finitude.

A Critique of "Substance"

The observations made in the previous section remain an empty exercise if one does not follow the path of Heidegger's engagement with Descartes' ontology to its limit. According to Heidegger, the clue to how spatiality is fixed into its interpretation as the constant, present spatiality of objective beings lies in the way in which Descartes deals with the double meaning of "substance." The use of this term points to an ambiguity in the way Descartes understands being. Substance can refer to a thing or an element (as in chemistry, for example), and at the same time, it can be used to refer to the being of that same thing or element, i.e., as the "substance" of this or that entity. For instance, "the space of nature" refers to the place of things as well as to the world-space "in" which all beings are encountered. One finds that with this ambiguous way of speaking about the spatiality of the occurrences of beings an opening appears, and one glimpses an abyss that remains the silent ground of Descartes' conception of substance by being covered over as a result of his focus on objective and ideal presence. This ambiguity concerning spatiality raises the question of how Descartes understands the being of substance. In terms of the tradition one may say that substance as a thing is grounded on substance as the ever-present, unchanging attribute of all beings. But what is the being of this attribute as substance, i.e., as such? Or, as Heidegger puts it, what is the "substantiality" of "substance"?

In §20 Heidegger points out that Descartes' conception of substance goes back to the substance that needs nothing else to be, the substance that is always and unchanging, since it is eternal.[48] This was also evident

48. *BT*, 86 (*SZ*, 92).

in the discussion of Descartes' "world." But that the idea of unchanging and ever-present substances has its roots in the idea of God presents an insurmountable problem for Descartes. How can one speak univocally of eternal (i.e., not created) being and of what is created? How can one bridge the infinite difference between God and the two ontological elements, *res cogitans* and *res extensa,* without reducing eternal being to the character of the created? As Heidegger shows, Descartes evades the question altogether by stating that what is known both to the Eternal and to beings cannot be known to the human intellect, to the created, to the *res cogitans*: "He actually evades the question. '*Nulla eius "substantiae" nominis significatio potest distincte intelligi, quae Deo et creaturis sit communis.*'"[49]

Heidegger then demonstrates that Descartes both avoids the question of the being of substance, and puts it out of the reach of thinking. "Descartes not only evades the ontological question of substantiality altogether; he also emphasizes explicitly that substance as such—that is to say, its substantiality—is in and of itself inaccessible from the outset. . . . Being does not affect us and therefore cannot be perceived. Being is not a predicate says Kant, who is repeating Descartes' principle. Thus the possibility of a pure problematic of being gets renounced in principle."[50] According to Descartes, substance is imperceptible, it is unknowable in itself. Heidegger points out that this amounts to rejecting in principle the question of being, and that this objection sets a precedent for the tradition of modern philosophy, for example, for Kant, who repeats this gesture in the *First Critique* by stating that being is not a predicate.[51]

That the being of substance is not knowable to human intellect does not prevent Descartes from formulating a distinct idea of it. The being of substance cannot be sought, but both the substance that makes up each particular and its attributes are accessible to pure reason. Thus, substance is understood in relation to the particular unchanging characteristics of entities present at hand. At the same time, these characteristics are neither perceivable by the senses nor encountered directly in our experience.

49. *BT,* 86 (*SZ,* 93). The quote is from Descartes, *Principles,* I.51 (VIII: a, 24 [210]).

50. *BT,* 126–27, M&R translation (*SZ,* 94).

51. This principle of exclusion is, of course, even more dramatically exercised by the interpretation of Kant's *Critique* as an inquiry into possible empirical knowledge within the limits of the rules and principles of reason. Such an interpretation takes the *Critique* to be an epistemological work concerned not with the question of being, as Heidegger claims in *Kant and the Problem of Metaphysics,* but with the transcendental condition for the possibility of Newtonian physics and scientific mathematical sciences, and empirical knowledge in general. Cf. *BT,* 2–3 (*SZ,* 3–5).

(Attributes are intuited by consciousness and made clear by the process of pure reasoning.)[52] In other words, Descartes abstracts from the phenomena that which he finds agreeable with his ungrounded idea of substance as constant presence. These characteristics of things he understands to be ever-present, unchanging principles of "nature," and as intuited substances, which he then employs to interpret the phenomena. On the one hand, Heidegger's critique shows how Descartes' thought closes on itself and gives all beings determinations in terms of objective and ideal presence. On the other hand, this interpretation of the grounds of Descartes' ontology shows how from the "space of nature" to the core of its ontologico-conceptual ground—that is, spatiality as substance, the spatiality of being—interpretations are always made in terms of the categorical way of being of objective presence. Ultimately, the concept of the "space of nature" occurs as an interpretation of the phenomena of the world or the occurrences of beings based on an ungrounded principle of constant presence.

Twisting Free

Heidegger's critique of substance in Descartes' thinking leads to the twisting free of spatiality from metaphysical and transcendental interpretations grounded on objective and ideal presence. In "A Hermeneutical Discussion of Descartes' Ontology," Heidegger demonstrates, as shown above, how Descartes' ontology is not only insufficient in its grounding, but that it "passes over" the phenomena of the world and the being of the entities found "in" the world.[53] In this section Heidegger argues that Descartes' insistence on the concept of substance and its constant presence leads him to prescribe "true" being to beings rather than letting the phenomena of the world show themselves. As indicated before, Heidegger writes that Descartes "rather prescribes to the world, so to speak, its 'true' being on the basis of an idea of being (being = constant objective presence) the source of which has not been revealed and the justification of which has not been demonstrated."[54] Descartes' emphasis on constant presence and immutability leads him to identify mathematical knowledge with substance.[55] Thus, the conceptuality and logic of substances dictate

52. Descartes, *Rules for the Direction of the Mind,* I, II, and III (X: 359–70 [9–15]).
53. *BT,* 88 (*SZ,* 95).
54. *BT,* 89 (*SZ,* 96).
55. Ibid.

the being of beings. In this way the phenomena of the occurrences of beings are interpreted as "nature" and in terms of "principles and laws." Not only does this mean that Descartes' question and reasoning make his ontology insufficient, and that his ontology misses or passes over the phenomena themselves, the being of the world; it thus never reaches either the phenomena or the spatiality of being. Ultimately, the phenomena are interpreted and judged in their being through these insufficient concepts and remain lost under such ruling conceptual determinations.

When he writes of Descartes' prescriptive mistake, Heidegger critiques Descartes not insofar as he fails to give an accurate conceptual account of the phenomena of world; the problem is that the concept of substance is "an idea of being whose source has not been revealed." In other words, the problem is that the preconscious event of the disclosedness of being, being-in-the-world, has not been revealed, let alone developed out of fundamental temporality. This criticism points back to the phenomena of the events of beings and their spatiality, i.e., the phenomena of the characteristic spatiality of being-in-the-world that remain covered over and passed over in Descartes' understanding of spatiality as "natural" and "substantial." At the same time, these phenomena lie beyond Descartes' ontology, and therefore can neither be attributed to the "space" of things, *res extensa,* nor to the intuition of consciousness, *res cogitans.* This is because, as the critique has shown, the grounding principle for these two elements of Descartes' ontology (*substantia*) does not reach phenomena of events of beings. From this it follows that, since the spatiality of the world remains beyond Descartes' ontology, the judgments and rules that gave it determination as *extensio* in terms of that ontology are no longer sufficient. Thus, spatiality has been set free, has twisted free from traditional metaphysics and its judgment of beings. This does not occur through Heidegger's critique of the conceptual structure of Descartes' ontology, but by his focusing on the phenomena of the world or the occurrences of beings as that which has been passed over and which remains to be sought as such.[56] But how is one to understand the matter that has been uncovered and opened for a new engagement?

As a result of Heidegger's critique, the twisting free of spatiality cannot point to another kind of being to be thought in the sense of an entity,

56. When Casey points out the dualism in Descartes' concept of "place" by introducing the difference between "internal place," the place attributable to *extensio,* and "external place," or relational space, he is pointing to a sense of spatiality outside of its traditional interpretation. See note 25.

concept, or substance to be grasped by consciousness, nor can it be situated in a self-evident consciousness. Heidegger's critique of the concept of substance eliminates these possibilities. Therefore the question that one faces now is not one asking what kind of being is space, nor is it a question of transcendental substances. Heidegger's critique points to the phenomena of the occurrences of beings in their temporality and finitude. At the end of his critique he writes that "[t]aking *extensio* as the basic determination of the 'world' has its phenomenal justification."[57] What is salvageable is not Descartes' interpretation of spatiality, but the phenomena of spatiality itself in all events of beings.[58] It is this indication toward the recovery of the spatiality of events of beings that introduces Heidegger's own discussion of the spatiality of dasein's being-in-the-world.

Conclusion: The Alterity of Critique and the Spatiality of Events of Beings

Heidegger's critique of traditional ontology is an attempt to recover the phenomena of the spatiality of beings in their temporality and finitude. But what is it that sustains Heidegger's critique, if not the assertive concepts of the very ontology he criticizes? In light of his refutation of objective and ideal presence as a ground for the interpretation of being, Heidegger's thought, his critique and recovery of the phenomena of world and spatiality, cannot be traced back to the ever-present and sustaining grounds of traditional ontology. How is one to understand the event of Heidegger's thought then? Up to now this discussion has followed Heidegger's arguments, but in order to engage this last question one must take up the performative aspects of his thought, and therefore move beyond Heidegger's explicit analysis of Descartes' ontology.

The release of spatiality from the metaphysical and transcendental tradition and the move toward the recovery of spatiality by another thought both initiate a radical event of alterity—the alterity of a thought that cannot trace its event to metaphysical, ever-present origins or to an ever-

57. *BT,* 94 (*SZ,* 101).

58. This is an issue that must be sought ontologically, and that is thematically engaged in *Being and Time* in terms of the spatiality of dasein's being-in-the-world.

present, transcendental structure, and that remains on the way toward the phenomena. Furthermore, this alterity indicates the exilic character of Heidegger's thought. On the one hand, Heidegger's critique of traditional ontology cannot be grounded on objective and ideal presence. On the other hand, as an indication of or negative introduction to a positive discussion of spatiality beyond the tradition, the critique can only look forward to a thought to come. Heidegger's critique therefore enacts an exilic disclosure of the spatiality of beings by occurring in a manner that does not refer the phenomena back to unchanging, ever-present origins, and in a manner that remains open to events that are yet to come. The question is now if and how Heidegger engages spatiality, as well as the alterity of thought and of the events of beings figured by this issue.

4

Failure, Loss, Alterity

Being and Time and Spatiality

Introduction

As the last chapter indicated, Heidegger's critique of Descartes' ontology releases spatiality from its traditional interpretations in terms of objective and ideal presence, and makes a move toward the recovery of the phenomena of the world and spatiality. This happens in light of a certain exilic aspect inherent in the very event of Heidegger's thought in that critique—i.e., in that after Heidegger's critique neither the metaphysical nor the transcendental grounds of the tradition can be taken as the basis for Heidegger's thought, and in that the engagement of the phenomena remains to be accomplished. Beginning, then, from Heidegger's critique of Descartes' conception of the world and spatiality, one sets out from these exilic grounds. But this situation points already to at least two necessary tasks for the engagement of the phenomena of world and spatiality: to engage the phenomena of world and spatiality in their temporality or finitude, i.e., in their concrete events; and to engage the spatiality or "taking place" of that very thought that engages the phenomena. As this chapter indicates, the shadow of alterity appears behind Heidegger's attempt to take up these tasks.

The impossibility of grounding Heidegger's thought on metaphysical or transcendental terms calls into question the character of Heidegger's positive account of spatiality: How does one begin to engage spatiality if not in terms of the tradition? A possible path is suggested by a warning born out of Heidegger's single focus on temporality in *Being and Time.* Even the most careless look at the book makes immediately apparent Heidegger's focus on temporality as "the" horizon and originary character of the occurrences or events of beings. Furthermore, this focus indicates a separation between *Being and Time*, the book about "being" and "time," the book of temporal ontology, and a difficulty brought to it, i.e., the issue of spatiality. The relentless focus on temporality as "the" horizon of the disclosedness of beings makes clear that spatiality comes to *Being and Time* as a foreign issue since, according to Heidegger, spatiality will only be understood and hence accepted into his discourse once temporality has been recovered. This sense of strangeness points first of all to the traditional interpretation of the work as needing to be engaged purely in terms of the question of temporality and being, an interpretation already operative by the time one begins to attempt to engage the difficult issues of spatiality as they appear in the book. But this sense of the strangeness of spatiality may also serve as a warning against the danger of beginning in a way that has already given up spatiality in the name of temporality. Indeed, this issue of spatiality as somewhat foreign to the main point of the book points to another difficulty, one that, rather than obscuring or concealing the issue of the spatiality of beings, serves as an engaging intimation of the difficulty of spatiality as operative in *Being and Time.* If spatiality comes to being as a stranger, it comes as a stranger on its way toward its native land. As we have already seen, in spite of Heidegger's focus on temporality, spatiality does not appear as an issue exterior to the question of being; spatiality is operative in the disclosedness of events of beings. Thus, if the ontological drive that seems to keep spatiality at a distance from Heidegger's analysis of the disclosedness of events of beings is to recall the question of being, this same drive must engage the originary being of spatiality operative in those events of disclosedness. Precisely in the difficulty of this "homecoming," in the difficult proximity of spatiality and being, does one begin to encounter spatiality in *Being and Time.*

In Heidegger's book the recalling of the question of being occurs through the ontological analysis of dasein, first by assessing its being-in-the-world in its everydayness, then by rethinking this structure in its fundamental temporality. In §70 of *Being and Time,* a latter part of the analysis of dasein, Heidegger attempts to "trace back" (*zurückführen*) dasein's existential spatiality to its temporality. In what follows I discuss how this is a failed attempt to think being's spatiality: Heidegger fails to engage the phenomena of spatiality because he articulates it transcendentally. This "failure" is manifested in the slipping away or withdrawal of spatiality from Heidegger's transcendental articulation of the question of being. This withdrawal from Heidegger's apophantic logos recalls the alterity of events of beings, while at the same time indicating the withdrawal of spatiality from Heidegger's language, a withdrawal that figures the alterity of that thought. In light of the withdrawal of spatiality, Heidegger's discourse cannot articulate its own event or "taking place." This withdrawal leads Heidegger to encounter the need for another way of engaging being, and it therefore marks a transformative passage in which Heidegger's thought leads beyond its conceptual determinations and articulation of the question of being in *Being and Time.* In its engagement of Heidegger's thought in its failure, this chapter to a certain extent takes up what is traditionally assumed to be meaningless by those assertive discourses grounded on objective and ideal presence: it engages failure as an operative aspect of Heidegger's discourse. This occurs as the discussion makes apparent that, in that failure or loss in the attempt to articulate spatiality in the question of being, thinking is recalled to the withdrawal essential to the occurrence of thought and of beings—i.e., to the alterity of thought and occurrences of beings. Furthermore, it is also this failure that brings forth the powerful transformative force of thought, since, as I have indicated above, this difficulty figures Heidegger's realization of the need for another way of engaging events of beings, a realization that will lead his thought in *Being and Time* beyond itself and that therefore will have placed thought on exilic grounds.

The Import of Spatiality: Dasein's Spatiality and the Spatiality of the Events of Beings

The matter of the spatiality of the occurrences or events of beings is directly raised in *Being and Time.* In §24, referring to his discussion of dasein's existential spatiality in this and the previous two sections, Heidegger writes, "[o]ur problematic is merely designed to establish ontologically the phenomenal basis upon which one can take the discovery of pure space as a theme of investigation, and work it out."[1] Heidegger's concern in his analysis of spatiality is not any particular "space," but the originary event that grounds spatiality. Spatiality is engaged here in terms of this originary event. But what is the sense of spatiality Heidegger is indicating?

As already noted in Chapter 2, spatiality in *Being and Time* does not refer to objective space, nor to the ideal or transcendental ideas based on entities objectively present at hand. The statement above goes further than this: Heidegger's analysis of spatiality precedes "the discovery of pure space as a theme of investigation."[2] It refers to an issue prior to the understanding of spatiality as homogeneous and intuited by consciousness. The term "pure spatiality" refers to Kant's understanding of "space" in the *Critique of Pure Reason* as the intuition of homogeneous space by a transcendental consciousness.[3] It also refers to Husserl's understanding of spatiality as intuited. In a footnote immediately preceding the passage in question Heidegger cites the work of his colleague and contemporary Oskar Becker, who in his book *Beiträge zur phänom-*

1. *BT,* 104 (*SZ,* 112).
2. Ibid.
3. In the "Transcendental Aesthetic," in the *First Critique,* Kant identifies two pure intuitions, time and space: the first corresponds to inner sense, and the second to outer sense. These two intuitions are pure (*rein*) in that they are a priori, i.e., they occur prior to and are the condition for the possibility of all sense experience. In the "Transcendental Exposition of the Concept of Space," Kant writes that "[f]urthermore, this intuition must be a priori, that is, it must be found in us prior to any perception of an object, and must therefore be pure (*rein*), not an empirical, intuition." In the "Metaphysical Exposition of the Concept," Kant states that "space is represented as an infinite given magnitude." He then concludes that "for all the parts of space coexist ad infinitum. Consequently, the original representation of space is an a priory intuition." Kant adopts Euclid's idea of space as continuous and homogeneous in his deduction of spatiality as a pure intuition. Immanuel Kant, *Critique of Pure Reason,* tr. Norman Kemp Smith (New York: St. Martin's Press, 1965), 65–78. See also *Kritik der reinen Vernunft* (Hamburg: Felix Meiner, 1956), 63–93.

Kant's understanding of space oscillates between the Newtonian version of infinite, absolute, uniform, and isotropic space, which leads his discussion in the *First Critique,* and his earlier understanding of space in terms of the work of Leibniz, for whom it is a relation subject to the simultaneously compossible states of monads—i.e., for Leibnitz there is no absolute space without substances in their relation. On this oscillation in Kant's work, see Lewis White Beck, *Early German Philosophy: Kant and His Predecessors* (Bristol: Thoemmes Press, 1969). See also *Historisches Wörterbuch der Philosophie,* ed. Joachim Ritter and Karlfried Gründer (Darmstadt: Wissenschaftliche Buchgesellschaft, 1992), 88–94.

enologischen Begründung der Geometrie und ihrer physikalischen Anwendung works out Husserl's understanding of space constituted through a transcendental reduction.[4] The footnote acknowledges Becker's work for disclosing the prescientific spatiality that precedes Euclidean forms of space. But this citation also places Becker's work, and through it Husserl's understanding of spatiality, in direct relation to Heidegger's own analysis.[5] Becker's analysis is set against the more fundamental task of Heidegger's analysis of existential spatiality. Heidegger's analysis precedes and exposes as inadequate the grounds for any transcendental reduction; it asks the question of the being of the consciousness that intuits spatiality (dasein), and simultaneously, in addressing the being of this consciousness, it aims to disclose the way being is already given in this event (being-in-the-world). The last point will disclose consciousness in its event or its temporality and finitude. As Heidegger points out in *Basic Problems of Phenomenology,* dasein's being-in-the-world is the grounds for intentional acts.[6] Thus, Heidegger's analysis will lead to the uncovering of the phenomenal grounds for traditional interpretations of spatiality.

When Heidegger speaks of the "phenomenal basis" of any discourse on spatiality, he refers to the characteristic spatiality of dasein's being-in-the-world, thought both in everydayness and out of originary spatiality.[7] In the same section in *Being and Time* in which he calls for the suspension of the objective interpretation of spatiality as "space," Heidegger adds, "[i]n the first instance it is enough to see the ontological difference between Being-in

4. *BT,* 104 n. 23 (*SZ,* 112 n. 1). Cf. Heidegger, *History of the Concept of Time: Prolegomena,* tr. Theodore Kisiel (Indianapolis: Indiana University Press, 1992), 324: "For those who are somewhat more thoroughly conversant with mathematical things, I refer to an investigation carried out within the purview of phenomenology by Oskar Becker. Here, to be sure, the essential question of the genesis of the specific mathematical space of nature from environmental space is not developed, although it stands in the background for the author."

5. Becker's work represents Husserl's understanding of space only at the point of his work on *Ideas,* after which Husserl's work will take a historical turn, as is the case in *Cartesian Meditations* and finally in the *Crisis of European Sciences.* See "Husserl on Space and Time," in *Husserl: Shorter Works,* ed. Peter McCormick and Frederick A. Elliston (Notre Dame: University of Notre Dame Press, 1981), 211–50.

6. *The Basic Problems of Phenomenology,* tr. Albert Hofstadter (Indianapolis: Indiana University Press, 1988), 161–73. Also *Die Grundprobleme der Phänomenologie, GA* 24, §15, 229–42: "Radikalere Interpretation der Intentionalität für die Aufklärung des alltäglichen Selbstverständnisses. Das In-der-Welt-sein als Fundament der Intentionalität."

7. One should keep in mind that there is a difference between Husserl's understanding of the *phainomenon* as the object of intentional acts of consciousness, and Heidegger's understanding of it as *phainesthai,* in its middle voice, as the coming into being of beings from themselves in their shining forth (i.e., in enactment).

as an existential and the category of the 'insideness' which things present-at-hand can have with regard to one another. By thus delimiting Being-in, we are not denying every kind of 'spatiality' to Dasein. On the contrary, Dasein itself has a 'Being-in-space' of its own; but this in turn is possible *only on the basis of Being-in-the-world in general*."[8] As we have noted, Heidegger explicitly calls for putting "out of play" the question of "space" in his *Prolegomena to the History of Time*, the earlier draft of *Being and Time*. But this necessary suspension is only sufficient as a "first instance," a first step toward the rearticulation of the spatiality of events of beings as disclosed in dasein (being-t/here). The suspension of the question of "space" serves as a moment of interruption and an orientation toward the engagement of spatiality. This suspension "keeps open the way for seeing the kind of spatiality which is constitutive for Dasein."[9] Heidegger's specific discussion of the spatiality of dasein's being-in-the-world occurs in Chapter 3, §C, immediately after his critique of Cartesian ontology.

Dasein's existential spatiality is an intrinsic element of the ontological constitution of its being-in-the-world. As the passage above indicates, "Dasein itself has a being-in-space of its own." The emphasis here should not fall on "having," as if spatiality were something other than dasein. The point is rather that spatiality is dasein's "own." According to Heidegger, dasein (being-t/here) occurs always as a spatial event. This is clear from Heidegger's breakdown of dasein's being-in-the-world into three parts, where "being-in" (*In-sein*) is one of its fundamental ontological elements.[10] The operative character of spatiality as one of dasein's aspects is formally indicated by the titles of §§23–24, "The Spatiality of Being-in-the-World" and "The Spatiality of Dasein and Space."[11] These sections consist of the concrete phenomenological analysis of dasein's own spatiality. In §23 Heidegger states that dasein's spatiality "is essentially de-distancing, that is, it is spatial."[12] And at the end of §24, Heidegger concludes the only general section on spatiality in *Being and Time* by stating that dasein is essentially spatial. "Indeed space is still one of the things that is constitutive for the world, just as Dasein's own spatiality is essential to its basic state of Being-in-the-world."[13] One may conclude,

8. *BT*, 82, M&R translation, italics in original (*SZ*, 56).
9. *BT*, 134, M&R translation (*SZ*, 101)
10. *BT*, 49–51 (*SZ*, 53–54).
11. *BT*, 94–105 (*SZ*, 101–13).
12. *BT*, 100 (*SZ*, 108).
13. *BT*, 147–48, M&R translation (*SZ*, 113).

then, from Heidegger's own words, that there is an existential spatiality that belongs to dasein's being-in-the-world as part of its ontico-ontological structure.

Indeed, dasein's spatiality leads beyond the space of any thing, entity, or subject. In *Being and Time* dasein's being-in-the-world figures the disclosedness of events of beings, as well as the spatiality of such events. This is distinctly indicated by Heidegger's discussion of dasein's being-in-the-world as the "clearing" (*Lichtung*); a term Heidegger will later radicalize but that already in *Being and Time* figures the disclosedness of events of beings, although in terms of their temporal horizon understood as dasein. In Chapter 3, §1 of *Being and Time,* Heidegger offers a thorough analysis of "being-in," the theme used to introduce dasein's ontological difference from objective presence in §12.[14] At the beginning of this chapter Heidegger not only repeats that dasein's being-in does not refer to the objective space of an entity at hand, but extends this earlier differentiation, maintaining that dasein's being-in is itself the opening of the disclosedness of being.

> The entity which is essentially constituted by Being-in-the-world is itself in every case its "there" [*Da*]. According to the familiar signification of the word, the "there" points to a "here" and a "yonder." The "here" of an "I-here" is always understood in relation to a "yonder" ready to hand, in the sense of being toward this "yonder"—a being which is de-severed, directional and concern-full. Dasein's existential spatiality, which thus determines its "location," is itself grounded in being-in-the-world. The "yonder" belongs definitely to something encountered within-the-world. "Here" and "yonder" are possible only in a "there"—that is to say, only if there is an entity which has made a disclosure of spatiality as the being of the "there." This entity carries in its ownmost being the character of not being closed off. In the expression "there" we have in view this essential disclosedness. By reason of this disclosedness, this entity (*Dasein*), together with the being-there of the world, is "there" for itself.
>
> When we talk in an ontically figurative way of the "*lumen naturale*" in man, we have in mind nothing other than the existential-ontological structure of this entity, that it is in such a way as to be

14. *BT,* 49–56 (*SZ,* 52–60).

> its "there" [*Da*]. To say that it is illuminated [*erleuchtet*] means that as being-in-the-world it is cleared [*gelichtet*] in itself, not through any other entity, but in such way that it is itself the clearing [*Lichtung*]. Only for an entity which is existentially cleared in this way does that which is present at hand become accessible in the light or hidden in the dark. By its very nature, Dasein brings its "there" along with it. If it lacks its "there," it is not factually the entity which is essentially Dasein; indeed, it is not this entity at all. Dasein is its disclosedness.[15]

In this passage, Heidegger points out that dasein occurs as its own disclosedness, as the "clearing" (*Lichtung*), and that it is out of this event of disclosedness that beings may show themselves in their clarity and darkness. To engage dasein's spatiality is to engage the spatiality of events of beings in their disclosedness. A situation analogous to dasein's being as *Lichtung* is found in §12, where Heidegger discusses the sense of "being-in" or "inhabiting" that belongs to dasein by showing how being by (*bei*) one another and touching (*berühret*) are ways of being that are only possible for entities out of dasein's "being-there" (*Da-sein*). "An entity present-at-hand within the world can be touched by another entity only if by its very nature the latter has being-in as its own kind of being—only if, with its being-there [*Da-sein*] something like another entity can manifest itself in touching, and thus become accessible in its being-present-at-hand."[16] The "there" of dasein and the discussion both of "being by" and "touching" and of the word *Lichtung*—which in German customarily means a clearing in the woods or an opening—refer to the disclosedness of events of beings not only in terms of temporality and finitude but also with direct indications as to the spatiality of such event.

As the language in the passage on *Lichtung* indicates, the emphasis is on "light." Light is traditionally associated with reason. The term *lumen naturale* refers to the characterization of dasein as *animal rationale.* The light of reason, *lumen naturale,* is thought to be its definitive characteristic, in the sense of the ability of reason to imitate and represent nature's first principles. But although Heidegger's choice of *Lichtung* also refers to *Licht* (light), the word points to an entirely different way of engaging beings. In German *Lichtung* never means "clarifying," "making clear," or

15. *BT,* 170–71, M&R translation, italics in original (*SZ,* 132–33).
16. *BT,* 81, M&R translation (*SZ,* 55).

"understanding through reason" (*ratio*), as in the light associated with *lumen naturale.* When Heidegger speaks of the disclosedness of events of beings as the *Lichtung,* he refers to a prerational disclosure. This differentiation excludes all interpretations of spatiality as a measurable and quantifiable event. The differentiation of *Lichtung* from *lumen naturale* raises the issue of the spatiality of the disclosedness of events of beings beyond the interpretation of spatiality as "space," identified with measurable extension, nature, and consciousness.

Looking closely at this aspect of Heidegger's engagement with the phenomenon of dasein's spatiality, it becomes clear that the issue is not the spatiality of a certain entity at hand but the spatiality of the disclosedness of events of beings. It is not from beings as entities that Heidegger's engagement with spatiality will come; rather the path of his thought is guided by an ontological concern. For Heidegger the question of the disclosure of being is an "ontological question." His understanding of "ontology" is directly stated in *Basic Problems of Phenomenology,* the seminar version of the unpublished third section of Part I of *Being and Time.* "Philosophy is not a science of beings but of being. . . . Philosophy is the theoretical conceptual interpretation of being, of being's structure and its possibilities. Philosophy is ontological. In contrast, a world-view is a positing knowledge of beings and a positing attitude toward beings; it is not ontological."[17] Philosophy and ontology are for Heidegger one single attempt to think being.[18] Thus ontology must think the being of beings out of being. Heidegger's way is not to proceed from entities toward being, rather it is always an attempt to think out of the event of being, out of the grounding question, the question of being.

In *Being and Time* Heidegger attempts to think the disclosedness of being by following the ground question of philosophy, "the question of being" (*Seinsfrage*). He seeks to reach this question through an analysis of dasein, the being that already has as its way of being an understanding (preconsciousness) of this question. Thus, he proceeds to uncover the question of being by giving a phenomenological analysis of dasein's everydayness, of being-in-the-world, by an analysis of the phenomenon of dasein's factical being. This occurs in Heidegger's phenomenological

17. Heidegger, *Basic Problems of Phenomenology,* 11. *Grundprobleme der Phänomenologie, GA* 24, 15.

18. Cf. Heidegger, "The Grounding Question and the Guiding Question of Philosophy," in *Nietzsche, The Will to Power as Art,* vol. 1, tr. David Farrell Krell (San Francisco: Harper, 1991); and Heidegger, *Nietzsche,* vol. 1, (Pfüllingen: Neske, 1989) 79–81.

approach to the formal structure of dasein's everydayness in Part I of *Being and Time,* including his formal discussion of the spatiality of the phenomenon of dasein's everydayness (in Chapter 3, §C). In Part II of the book Heidegger takes these formal elements that he has uncovered in his phenomenological analysis and attempts to engage them in terms of their originary temporality. He attempts to think spatiality out of temporality in §70 of Part II of the book, entitled "The Temporality of Dasein's Characteristic Spatiality."[19]

As is well known, the third part of the book was never published. Heidegger's project extended to this third part, in which, in light of the articulation of dasein's temporality, he intended to think beings out of being, out of their finitude or temporality—i.e., in their coming to be in passing away. This emphasis on "ontology" behind Heidegger's project must be clear because it implies a certain direction in his thinking. The motion of Heidegger's thought is not from beings (entities at hand) to their unchanging principles, but from an intimation of being toward the articulation of this intimation in its disclosive character, i.e., the analysis of dasein's preconceptual understanding of being in its way of being-in-the-world. Thus, in the dasein analysis, being (temporality) is not only the goal, but it is the grounding question for the analysis of being-in-the-world and this event's spatiality. (Heidegger makes this explicitly clear in his discussion of the structure of questioning in general, in the first part of the introduction to *Being and Time.*)[20] When Heidegger discusses spatiality in *Being and Time* he is trying to do so out of the grounding question, out of being (understood as the temporal horizon of the occurrences or events of beings). This also means that in *Being and Time* Heidegger attempts to think spatiality out of being. It is crucial to keep this in mind, since Heidegger's discussion of spatiality takes its departure from this single ontological direction.[21]

19. *BT,* 335 (*SZ,* 367).

20. *BT,* 4 (*SZ,* 5): "The Formal Structure of the Question of Being."

21. This is not to say that *Being and Time* is to be forced together with Heidegger's thinking in the 1930s. But, as Heidegger himself points out, the thinking of *Being and Time* is already essential and elemental to the turning. "Since the spring of 1923 the plan has been firmly established in its main features and it achieves its first shaping in the projecting-opening called 'From Enowning.' Everything moves toward this projecting-opening, including *Eine Auseinandersetzung mit Sein und Zeit,* which also belongs to the domain of these mindful deliberations. *Beilage zu Wunsch und Wille: Über die Bewahrung des Versuchten.*" *Besinnung, GA* 66; *Eine Auseinandersetzung mit Sein und Zeit* (Frankfurt am Main: V. Klostermann, forthcoming).

The Transcendental Articulation of the Spatiality of Being

Although Heidegger maintains in Part I of *Being and Time* that dasein's being-in-the-world is characteristically spatial, when he attempts to rethink the phenomena of spatiality and interpret it in terms of temporality he falls into a transcendental articulation that pushes spatiality away from the question of being. Heidegger's attempt to think spatiality out of its temporality occurs in §70 of Part II, "The Temporality of Dasein's Characteristic Spatiality."[22] The title indicates the direction of Heidegger's thought, referring to the temporality "of" spatiality; it holds the two terms together, and this suggests a time-space play. At the same time, the discussion is meant to occur from out of being, since the leading question is that of temporality, and it is clear that Heidegger is attempting to think the spatiality of being in its event or passage.

At the beginning of this section Heidegger sets up the analysis of dasein's existential spatiality and originary temporality.

> Thus, with Dasein's spatiality, existential-temporal analysis seems to come to a limit, so that this entity that we call "Dasein," must be considered as "temporal" and also as spatial coordinately. Has our existential-temporal analysis of Dasein thus been brought to a halt by that phenomenon with which we have become acquainted as the spatiality that is characteristic of Dasein and which we have pointed out as belonging to being-in-the-world?
>
> If in the course of our existential interpretation we were to talk about Dasein's having a "spatio-temporal" character, we could not mean that this entity is present-at-hand "in space and also in time"; this needs no further discussion. Temporality is the meaning of the being of care. Dasein's constitution and its ways to be are possible ontologically only on the basis of temporality, regardless of whether this entity occurs "in time" or not. Hence Dasein's specific spatiality must be grounded on temporality. On the other hand, the demonstration that this spatiality is existentially possible only through temporality, cannot aim either at deducing space from time or at dissolving it into pure time.[23]

22. *BT,* 335 (*SZ,* 367).
23. *BT,* 418, M&R translation (*SZ,* 367).

Heidegger's beginning shows that he is clearly aware of the difficulty of giving a purely temporal analysis of the disclosedness of events of beings in light of the existential spatiality that belongs to dasein's being-in-the-world. Because of the spatiality of dasein's being-in-the-world it seems that dasein will have to be considered not only temporally but also spatially. Heidegger makes a path toward this articulation by saying that spatiality must be engaged out of dasein's being, i.e., out of its originary temporality: "Dasein's specific spatiality must be grounded on temporality." Heidegger points out that spatiality will not be engaged by referring to it as a "spatio-temporal" event because such an articulation is a return to the understanding of spatiality in terms of the objective interpretation of spatiality and temporality. So, he writes, "we could not mean that this entity is present at hand 'in space and in time.'" Of course, this does not make a claim for the separation of spatiality from temporality in the disclosure of being. The sentence only points out that the spatio-temporal disclosedness of the occurrences or events of beings should not be sought in terms of objective and ideal time and space. Following the ontological direction of his path, and in order to orient his inquiry, Heidegger then makes a leap from the objective sense of being into being. He continues, "[t]emporality is the meaning of the being of Dasein's care." In making such leap Heidegger attempts to articulate dasein's existential spatiality out of pure temporality, i.e., out of being. This introductory sentence is followed by Heidegger's attempt at an existential analysis of the temporality of dasein's existential spatiality. But in the leap to begin thinking the spatiality of being out of being, Heidegger severs dasein's existential spatiality from its temporality. He writes that "Dasein's specific spatiality must be grounded on temporality." The "grounding" of dasein's spatiality in temporality seems to draw a line, to mark a limit between dasein's being and dasein's temporal origin; it also seems to draw a "horizon" of temporality that one must reach by moving out of dasein toward temporality through the dasein analysis.

The full import of this moment strikes when one views it in the context of Heidegger's project as a whole. As already indicated, the aim of Heidegger's project in *Being and Time* is "ontology," i.e., to think the being of beings out of being as such.[24] The separation of existential spatiality from temporality is a transcendental move. Heidegger writes in this same passage, "[t]emporality is the meaning of the being of care.

24. See notes 17 and 18.

Dasein's constitution and its ways to be are possible ontologically only on the basis of temporality, regardless of whether this entity occurs 'in time' or not." It seems here that events of beings, including dasein's existential spatiality, are ontologically possible on the basis of temporality alone. Indeed, in light of the separation of dasein's existential spatiality from temporality, Heidegger's remark above hints at a pure temporality unblemished by the concreteness of the finitude of dasein and of the occurrences or events of beings.

This transcendental aspect of Heidegger's articulation of the disclosedness of beings can be explored further by returning to the section in which Heidegger introduces the *Lichtung*. There, Heidegger writes that "the existential spatiality of Dasein, which in this way determines its 'place,' is itself grounded on being-in-the-world."[25] In other words, spatiality is "grounded" on the *Lichtung*. This suggests that being-in-the-world is a twofold event. At the level of facticity it is spatial, while at the disclosive level it is not. First of all, dasein's transcendental articulation turns temporality into an originary precedent over all beings. As such, temporality becomes a pure element of disclosure out of which beings are possible. It also suggests that the *Lichtung*, seen from the viewpoint of its temporality as the horizon of being, may be interpreted as a kind of deep phenomenon, as the more originary way of being that is the condition for the possibility of dasein's facticity and its spatiality. Interestingly, one can follow this a little further and conclude that being is not spatial, since the disclosure of being in its "pure" temporality is itself the condition for the possibility of spatiality. Of course, Heidegger's emphasis on temporality in *Being and Time* only confirms this reading. But this is not quite sufficient as an interpretation of the matter when one recalls that dasein is ontico-ontological, which translates into the inseparability of dasein's factical and ontological being, and thus into the inseparability of spatiality from temporality. Furthermore, the same inseparability of beings and dasein's events can be seen in dasein's being-in-the-world, which can only occur with dasein's (being-t/here's) being in the open with beings. In this sense, the possibility of dasein or the *Lichtung* can only be found in events of disclosedness in the plurality, diversity, and unique concreteness

25. I have retranslated this part of the passage to make more explicit Heidegger's emphatic separation between existential-spatiality and its ground. "Die existenziale Räumlichkeit des Daseins, die ihm dergestalt seinen 'Ort' bestimmt, gründet selbst auf dem In-der-Welt-sein." *BT*, 125 (*SZ*, 132).

of beings in their finitude (understood ontologically and not as entities with properties and substances).

The separation between dasein and temporality in the articulation of being-in-the-world also results in a characterization of dasein as an entity that must transcend itself in order to get to the horizon of being, to its authentic and ownmost modality of disclosure. Again, this is contradicted early on by Heidegger's text, when he declares that dasein is always both authentic and inauthentic. "As modes of being, authenticity and inauthenticity (these expressions have been chosen terminologically in a strict sense) are both grounded in the fact that any Dasein whatsoever is characterized by mineness. But the authenticity of Dasein does not signify any 'less' being or any 'lower' degree of being."[26] Indeed, as we have seen in various ways, the difficulties found in Heidegger's transcendental articulation of the disclosedness of beings can be distinguished from his project. But the difficulties brought forth by the points discussed are essential to the thought of *Being and Time.*

His transcendental articulation of the question of being in *Being and Time* is precisely the problem Heidegger explicitly confronts in his later work *On Time and Being.*[27] In this text Heidegger writes that "[t]he attempt in *Being and Time,* section 70, to trace back [*zurückführen*] Dasein's spatiality from temporality is untenable."[28] In view of Heidegger's separation of existential spatiality from dasein's being, the way to understand why Heidegger abandons his attempt to ground spatiality in temporality lies in how he tries to bring spatiality to being. Heidegger's word in describing his attempt to trace spatiality back to temporality is *zurückführen,* which literally means "to lead back." In *Being and Time* Heidegger is trying to lead existential spatiality back to its being, and in order to do so, he presents being as pure or originary temporality. But in doing so, he creates a transcendental horizon. In his attempt to trace spatiality back to being through pure temporality, he establishes a break between spatiality and being, and here the temporal horizon of dasein's being-in-the-world appears. This transcendental articulation is already announced in the first page of *Being and Time,* where Heidegger states

26. *BT,* 68, M&R translation (*SZ,* 43).

27. Heidegger, *On Time and Being,* tr. J. Stambaugh (New York: Harper & Row, 1972). See also "Zeit und Sein" and "Protokoll zu einem Seminar über den Vortrag 'Zeit und Sein,'" in *Zur Sache des Denkens* (Tübingen: Niemeyer, 1988).

28. My translation of "[d]er Versuch in 'Sein und Zeit' 70, die Räumlichkeit des Daseins auf die Zeitlichkeit zurückzuführen, läßt sich nicht halten." "Zeit und Sein," 24.

that "[o]ur provisional aim is the interpretation of time as the possible horizon for any understanding whatsoever of being."[29]

That Heidegger becomes aware of this transcendental and problematic aspect of the thinking of *Being and Time* is clear from his marginal comments to his own copy of the book. He writes in a marginal note to the title of the third section of Part I with regard to the dasein analysis, outlined at the end of the introduction, that in order to think being thinking needs "the overcoming of the horizon as such."[30] Instead what is called for is "[t]he return to the origin. The coming to presence out of this origin."[31] Heidegger points to such an articulation of spatiality in the passage from *On Time and Being* above.[32] To close, Heidegger "fails" in his attempt to think being's spatiality out of the intimation of the question of being, not through the reductive interpretation of the logic of categorical beings, but in giving a transcendental articulation to spatiality, an articulation that pushes spatiality away from being.

Slipping

A further look into Heidegger's differentiation between the event of existential spatiality and originary temporality shows that this ultimately leaves spatiality open to be thought. The transcendental structure of the move from objective space to pure temporality lets dasein's existential spatiality slip away from the analysis, since this spatiality belongs neither to objective space nor to being in its pure temporality. (As already indicated in Chapter 2, Heidegger's discussion of "being-in" in §12 makes clear that the event of dasein's existential spatiality cannot be thought in terms of objective space.) At the same time, the event of spatiality is kept separate from being by Heidegger's transcendental differentiation in §70. Dasein's existential spatiality is, therefore, an event that marks a place between the traditional transcendental elements of the understanding of being. Spatiality is neither attributable to beings nor to the transcendental articulation of being. In other words, dasein's existential spatiality does not have a place in Heidegger's analysis, or if it does, it is as a curious

29. *BT,* 1 (*SZ,* 1).
30. "Randbemerkungen aus dem Handexemplar des Autors." *BT,* 35 (*SZ,* 440).
31. Ibid.
32. *On Time and Being,* 24.

event between objective determination and originary temporality. This displacement is doubly enacted in §70: once in Heidegger's attempt to trace existential spatiality back to pure temporality; then a second time, retrospectively, in his abandonment of this attempt in *On Time and Being*, which indicates the difficulty and strangeness of the place of dasein's existential spatiality in the existential, temporal analytic of *Being and Time*.

Conclusion

The slipping of spatiality from Heidegger's transcendental articulation of the disclosedness of events of beings is a series of difficulties echoed by the figure of *chora* in Timaeus' likely story. First of all, in slipping away from Heidegger's articulation, spatiality figures a withdrawal (*choreo*), and much like *chora*, remains outside the philosophical discourse. Here, the spatiality of the disclosedness of events of beings has withdrawn, and therefore one finds the alterity of events of beings enacted in a coming to presence that figures a withdrawing motion. Also, as is the case with Timaeus' likely story, as well as with Aristotle's *logos apophantikos*, Heidegger's philosophical discourse travels to the limits of what can be said through discourse, and at that point finds itself recalled to its alterity by the withdrawal of being. It is not only that the disclosedness of beings has not been articulated, but that the event of articulation, the thought, also remains beyond articulation. What has withdrawn is the spatiality of the *Lichtung*, the spatiality of the occurrences or events of beings in disclosedness. But according to Heidegger this spatiality is none other than the spatiality of thought and language, since, as indicated in Chapter 2, the event of disclosedness in dasein is never other than language. Language and thought ultimately remain in alterity in light of the withdrawal of the spatiality of the very event of their articulation.

However, unlike the ancients, Heidegger's words do not simply leave their alterity to oblivion by asserting their re-presentation of being. Unlike Plato and Aristotle, Heidegger does not remain with an account of being, with the presence of the logos. In their return from the limits of language and in the enactment of the motion of his thought, Heidegger's words mark and engage their limit. The withdrawal that marks the "failure" of Heidegger's articulation not only indicates the alterity of thought, but also reveals the transformative force in the passage of that thought. Once again echoing the difficulty of *chora,* the withdrawal of spatiality figures the loss that occurs in the coming to presence of conceptual determinations, a loss operative in discourse. Unlike the tradition of representation, Heidegger's thought does not find meaninglessness or nothing in that withdrawal. Rather, in light of the withdrawal, Heidegger finds a call for another way of thinking, for another way toward the engagement of the occurrences or events of beings. This call is followed by Heidegger's refusal to publish §3 and Part II of *Being and Time,* his abandonment of §70 of the book, and, in its most dramatic form, in the famous "turn" (*Kehre*) after *Being and Time.* These are indications of the transformative passage Heidegger finds in the withdrawal of spatiality, and his touching on it.

Although Heidegger's articulation of the disclosedness of events of beings remains in the transcendental lineage when he separates dasein from pure temporality, out of this same motion of thought comes an interruption of this traditional interpretation of events of beings. In the slipping away of spatiality between objective space and pure temporality an interruption occurs in the transcendental interpretation of being. This is an event intimated in the withdrawal that remains beyond transcendental thought, an event that calls for a way of engaging the disclosedness of events of beings and their spatiality. From Heidegger's failure emerges a transformative force that calls for thinking both not out of ever-present being and toward its representation. Out of withdrawal, alterity, and loss, and in the collapse of a whole tradition of interpreting the occurrences or events of beings in terms of objective and ideal presence, thought will find a call for beginning anew from the events of the phenomena.[33]

The discussion has now moved from *Being and Time* beyond Heidegger, because the question of being-in-the-world is no longer safely contained by the horizon of pure temporality. If Heidegger's analysis aims not to be a story about beings,[34] it occurs as a double loosening up of the

33. Spatiality is not nothing; on the contrary, it is an irreducible phenomenon. Dasein's existential spatiality as the in-between marks a "place," an event. Although Heidegger speaks of "grounding" spatiality in temporality, at the same time he states in §70 that this spatiality can be neither "deduced" nor "dissolved into pure time" *BT,* 336 (*SZ,* 367). Dasein's existential spatiality is an irreducible event. It occurs literally as a "taking place" (*Raum einnehmmen*), which opens the clearing for beings in their spatiality (*einräumen*) and that, in doing so, simultaneously finds its own spatiality. This opening up occurs as a free-play for dasein. Dasein has the "leeway" (*Spielraum*) not only to give room for the shining forth of beings and their objective spatiality, but to also always open to its own place. It is this "taking place" that remains to be thought, not against being as its horizon, but out of such an event of being as such.

34. "If we are to understand the problem of being, our first philosophical step consists in not 'μυθον τινα διηγεισθαι,' in not 'telling a story.'" *BT,* 5 (*SZ,* 6).

tradition—double because not only is there no likely story, but also because the transcendental apophantic logos of time has released itself from pure temporality, and has been "contaminated" by the irreducible and ungraspable event of spatiality, which figures the alterity of the occurrences of beings and of thought. There will be no thinking that can simply attempt a return to temporality as the sole origin of all beings, as their possibility of disclosedness, without that thought's enactment of a certain alterity. Only if for a moment, in this space of withdrawal, loss, and transformation, the spatiality of thought is felt in its alterity. Now, one may return to the figure of spatiality as a stranger on its way "home." Spatiality never returns to temporality in *Being and Time.* However, the strangeness of the figure of spatiality, in its slipping, loosens up the transcendental tone of Heidegger's project and carries it beyond the transcendental thought of the book, toward an open path. Now what remains is the matter to be thought, an open difficulty arisen in withdrawal and alterity.

5

Enactments of Alterity

Heidegger's "Translation" of Spatiality

Introduction

Throughout this work spatiality appears as a figuration of the alterity of events of beings and events of thought. As such, the issue of spatiality bears the possibility of an opening toward a thought that will engage the occurrences of beings and their events and passages, in their alterity and on exilic grounds. In Chapter 1 the logos is discussed in terms of its limited function in Plato and Aristotle, i.e., as a mimetic tool for the representation of ever-present and unchanging ideas, essences, or first principles, a function that limits language to the service of entities present at hand and their ideal presence. As such, the logos appears at the limit of its representational power, as is indicated by the figure of *chora* in the *Timaeus.* In Chapters 3 and 4 this interpretation of language and philosophical discourse in terms of objective and ideal presence is interrupted, and along with this interruption, language and thought are caught in their alterity through philosophical discourse's failure to ultimately articulate its event. This occurs as the figure of the spatiality of the disclosedness of beings withdraws from discourse into concealment. Furthermore, in this withdrawal, discourse encounters the loss operative in its events, and with it, intimations of its alterity and exilic grounds. In the case of Plato and Aristotle, the emphasis on objective and ideal presence leaves unheard the intimation performed by such figure as *chora.* In Heidegger's case, in light of the withdrawal of the spatiality of the disclosedness of events of beings, philosophical discourse is caught in its alterity. However, rather than covering over alterity, Heidegger's thought remains engaged with the alterity of events of beings and of its event. This chapter concerns the alterity operative in Heidegger's thought in *Being and Time.*

The transformative motion of Heidegger's thought—thought's motion beyond its transcendental articulation of the disclosedness of events of beings and its critique and interruption of the tradition, in light of which thought cannot seek ground in objective and ideal presence for the understanding of its events—gives rise to a series of questions concerning his philosophical discourse. How is one to understand Heidegger's language in light of the interruption of the objective and ideal presence that traditionally gives meaning and form to language? How is one to engage Heidegger's language in a way that lets the alterity of that thought and of language be heard? In short, how will one engage thought in light of its alterity and exilic character? The aim of this chapter is to engage the passage or event of Heidegger's thought in its alterity and on exilic grounds. Once again this is done by taking up the figure of spatiality in *Being and Time.* This time the discussion will concentrate on Heidegger's analysis of spatiality in §§22–24. Again, the discussion follows closely both what Heidegger says and the motion or performative aspect of his thought.

Heidegger's discussion of spatiality in §§22–24 of *Being and Time* can be understood as an "ontological genealogy of space: how it arises in Dasein's world."[1] As such, it can be read as a "regrounding" of spatiality. This regrounding occurs immediately after spatiality has been severed from its traditional interpretations,[2] by Heidegger's critique of traditional ontology in the sections that immediately precede it. The general direction of *Being and Time* sustains the interpretation of these sections as a regrounding of spatiality, since the fundamental analysis of being-in-the-

1. Edward Casey, *The Fate of Place* (Berkeley and Los Angeles: University of California Press, 1997), 252.
2. *BT,* §§19–21 (*SZ,* §§19–21).

world and the existential structure of dasein's care, and ultimately its temporality, can be read as the preconceptual ground for metaphysics, the philosophy of nature, and transcendental philosophy.[3] But however powerful one's drive to "reground" spatiality, Heidegger's discussion must also be understood in terms of the critique of metaphysics and transcendental philosophy that introduces and precedes it. Therefore, in view of Heidegger's critique of traditional ontology, and because of his use of the language of this tradition,[4] his discussion of spatiality cannot be read but as a transformative appropriation, literally a "translation,"[5] of the traditional lineages and conceptualities that sustain interpretations of spatiality in terms of objective and ideal presence. The central issue here is the reappropriation of language as an "apophantic logos" in *Being and Time,* i.e., language understood as the possibility of letting beings appear as such, in their temporality and finitude—in light of their eventuation beyond objective and ideal presence alone.[6] The focus on language leads to the understanding of Heidegger's discussion as a double movement: on the one hand, it enacts the withdrawal of spatiality and the loss operative in the configuration of his thought. This is apparent in his narrow discussion of spatiality in terms of the necessary ordering for the presencing of beings as entities.[7] On the other hand, this thought also performs a movement that ultimately cannot be grounded either in traditional conceptuality or in Heidegger's own transcendental discussion of spatiality in *Being and Time.* Together, these aspects of Heidegger's discussion indicate the alterity and exilic grounds of his thought.

3. In §24, Heidegger writes that "[i]n our problematic we wish solely to establish ontologically the phenomenal basis for the thematic discovery and working out of pure space." *BT,* 104 (*SZ,* 112).

One should be attentive to how these readings of Heidegger's discussion sustain transcendental lineages by such questions as the ground of being and the condition for the possibility of being, if not by the question of substance (dasein as the authentic human).

4. In this discussion Heidegger uses terms like "place" (*Platz*), "space" (*Raum*), "region" (*Gegend*), and "de-distancing" (*Ent-fernung*).

5. This refers to "translation" in its literal Greek sense of changing (*metaphero*), "letting become otherwise," rather than, as traditionally understood, in terms of moving something unchanging from one point to another.

6. *BT,* 23–34 (*SZ,* 27–39).

7. "Presencing" is meant to guard us against the conflation of "ontic presence" and its "space" with the presencing and spatiality that concern Heidegger. For Heidegger presencing and presence (*An-wesenheit*) are to be understood neither ontically nor as the object of acts of an intentional consciousness (as would be the case with Husserl).

Translation

The difficulty in using a language dense with the traces and lineages of the logic of objective and ideal presence is clear for Heidegger from the start of *Being and Time.* He is certainly aware of it when, at the very beginning of his analysis of dasein's being-in-the-world, he calls for the suspension of spatiality when he introduces "environment" (*Umwelt*) and points out that the "aroundness" (*um*) at play in *Umwelt* does not have a primarily spatial sense.[8] Heidegger's sense of the difficulty of the articulation of being in language also appears throughout *Being and Time* in his struggle with language and with the temptation to return to an interpretation based on objective and ideal presence. Each time that he introduces such terms as "environment" or "region," Heidegger immediately warns that these terms are used otherwise than on the grounds of entities present at hand and the metaphysical and transcendental lineages that arise out of such engagement with the occurrences or events of beings. As he indicates, the terms he uses must be traced back to being-in-the-world, to the coming to presence of beings in their disclosedness with dasein. In other words, he must continuously pull thought back and distance it from the logic of objective and ideal presence.[9] This is not a mere call for "cleaning up" language. What is indicated is the difficulty of the task of thinking the spatiality of being in light of his critique of the tradition and his understanding of language as the possibility of the disclosedness of events of beings (as *a-letheia,* or a double motion of revealing and concealing).[10]

In his discussion of spatiality in §§22–24, Heidegger purposely uses the language that he has just rejected as ontologically and phenomenally insufficient with respect to the understanding of dasein's spatiality.[11] Some of his terms are: *Umwelt, Dort, Da, Wohin, Gegend, Ausrichtung,*

8. "The quality of 'around' (*um*), which is constitutive of the surrounding world (*Umwelt*), does not, however, have a primarily 'spatial' meaning." *BT,* 62 (*SZ,* 66). Cf. *GA* 20, 230; see also my discussion in Chapter 2.

9. Heidegger himself is led to wonder about the "distance" that is overcome in dasein's "de-distancing" (*Ent-fernung*). Concerning a passage on page 105 of *Being and Time* he writes later, in a marginal note in his working copy, "Where does the distance come from that is de-distanced? (*Woher die Ferne, die ent-fernt wird?*)" *BT,* 97 (*SZ,* 442).

10. *BT,* 23–34 (*SZ,* 27–39).

11. "We must show explicitly that Descartes not only goes amiss ontologically in his definition of the world, but that his interpretation and its foundations led him to pass over the phenomenon of world as well as the being of innerworldly beings initially at hand." *BT,* 88 (*SZ,* 95).

Raum, and *Räumlichkeit*. However, this language must now be reconsidered in light of his critique and call for a transformative reappropriation (*Destruktion*) of the tradition. Heidegger understands thought as a *Destruktion* (deconstruction) of the tradition: "We understand this task as the deconstruction [*Destruktion*] of the traditional content of ancient ontology which is to be carried out along the guidelines of the question of being."[12] This deconstruction begins with his critique of Descartes' concept of *substantia*, with the negative introduction to the issue of spatiality. This introduction, at the same time, brings the tradition to its originary event in the disclosedness of all occurrences of beings. Heidegger says as much when he calls for the analysis of Descartes' thinking out of being in the last section of his critique: "The following reflections can be grounded in more detail only by the phenomenological deconstruction [*Destruktion*] of the *cogito sum*." But this move is also enacted by Heidegger's transformative leap from the language of presence to a language that lets show the existential dimension of the spatiality of dasein's being-in-the-world in his discussion in §§22–24.

The use of traditional terminology in these sections indicates his attempt to leap from the traditional and reductive language of objective and ideal presence to a language that lets beings and their spatiality show themselves from themselves in their coming to be. But this task can only be accomplished through the transformative reappropriation of the language of the tradition. This is a crucial point for understanding Heidegger's discussion of spatiality: it occurs in full consciousness of the difficulty of rethinking the language of the tradition in a transformative engagement. Indeed, a close look at Heidegger's discussion of spatiality in §§22–24 will show that it occurs as a "translation" of traditional language. (As already indicated above, by "translation" I mean the literal sense of the word in its Greek root. "Translation" comes from the Latin *translatio*, which is a rendering of the Greek *metaphero*, literally meaning "to change," "to use a word in a changed sense.")[13] Heidegger's thought can be traced, then, by remaining with him in this difficulty with language, in this difficult entanglement of spatiality and traditional language.

But in what sense can the discussion be said to mark a reappropriation of the tradition? The terms Heidegger uses have been severed by this critique of the tradition from the structure of "meaning" in terms of objective

12. *BT*, 20 (*SZ*, 22).

13. *Metaphero* also means to "carry over," and the Latin term *translatio* conveys this narrow sense of the word, which is the common way of understanding our English term "translation."

and ideal presence. Therefore, that he uses them now already suggests that they are meant to be heard otherwise than in the tradition.[14] They must to a certain extent function otherwise than as traditional terms. When Heidegger remains with the language of the tradition, he interrupts the "meanings" that hold the tradition in place by finding new functions for the language that has sustained those meanings throughout the tradition. In using traditional language after it has been severed from its traditional interpretation, Heidegger is "re-turning" language,[15] bringing it to play otherwise than as traditionally understood. The language is decentered with respect to its meaning by Heidegger's critique of Descartes' ontology, and is then used outside of the traditional meaning. Here, although the play in language does reflect its hermeneutical context as it engages traditional lineages, it does so in light of the critique of the tradition; therefore, Heidegger's terms cannot be traced to an origin outside of the transformative leap he is making. This last observation also indicates that, in this re-turning of language, the lineages or ideas that sustain the interpretation of the occurrences of beings in terms of objective and ideal presence will no longer be operative in the engagement of beings.

When one looks at the position of Heidegger's discussion of spatiality in the general structure of the project of *Being and Time,* one sees that the discussion in §§22–24 focuses solely on dasein's facticity and the disclosedness of entities with factical dasein. The discussion of dasein's spatiality occurs in Part I, Chapter 3 of *Being and Time.* The aim of Part I is to provide a preparatory analysis of dasein's ontological structure, which entails the formal analysis of dasein's most proximate way of being-in-the-world.[16] In

14. Heidegger is not asking us simply to change from one "view" of spatiality to another, but to begin from a place utterly different from our sense of our surroundings, our body, "ourselves" as exterior and interior beings, as corporeal and spiritual. This is the translation taking place in §12, in the move from ontic "space," from the understanding of dasein as human body and spirit to its "inhabiting" (*wohnen*), its "being by" (*sein bei*) in the open with beings.

15. I point here to a dimension of Heidegger's language, not in the sense of Heidegger using a "tool" readily at hand for him, but in the sense of speaking in the possibility that is the essence of language, the possibility of the letting be of being. My use of "re-turn" and "re-turning" refers to this possibility by pointing to the play of language in its letting be, a play that is readily intimated in poetry. Poems occur as "verses." In "verse" what one literally find is a *versus,* a turning, the making of a turn. This observation also suggests the question of how Heidegger's famous "turn" (*Kehre*) is interpreted in relation to his prior thinking in *Being and Time.* Does it not make sense to ask about *Being and Time* and Heidegger's later thinking also in terms of this turning and "re-turning," in the sense of the openness of the possibility of language in its letting be?

16. *BT,* 62 (*SZ,* 66).

other words, its aim is to give a formal analysis of the preconscious comportment that reveals dasein's understanding of being in its everyday living, its facticity.

Chapter 3 falls within this larger project. It is a formal analysis of the world element of dasein's ontological structure (its being-in-the-world). The analysis focuses on dasein's everydayness, or its factical worldliness. In his introduction to this chapter, Heidegger refers to the world of factical dasein as "that 'in which' a factical Dasein lives."[17] The discussion of spatiality occurs as part of this chapter, and as such, it is concerned particularly with the factical phenomenon of dasein's everydayness. According to the general structure of the book, the discussion of the spatiality of dasein occurs as part of the preliminary uncovering of the ontological structure of dasein's being-in-the-world. The structure uncovered by this formal analysis of the phenomena of being-in-the-world still remains to be reconsidered in terms of its temporality or originary disclosedness in Part II of the book.[18] Heidegger's discussion of spatiality is preceded by his analysis of the "worldliness of the world."[19] In §15 Heidegger differentiates ontic beings from their being in their disclosedness with dasein. He refers to ontic beings as "being-at-hand" (*Vorhandensein*), and to beings in their disclosedness as being "handy" (*Zuhandensein.*)[20] The spatiality addressed in §§22–24 is thought out of the disclosedness of beings with dasein in their "handiness" (*Zuhandenheit*). Heidegger points to this by referring to these entities as "inner-*worldly.*"

The "translating" movement in Heidegger's thought is already apparent in §21, in his subtle translation of *Umwelt* from an ontic understanding of it, in terms of a measurable and local "surrounding" suggested by the "*um*" of *Umwelt,*[21] to an existential aspect, as the disclosedness of beings in their "being for the sake of," the "*um zu . . . ,*" and in light of dasein's finitude. Heidegger's discussion of the spatiality of the *Umwelt* is an attempt to think the spatiality of the "surrounding world" out of the disclosedness of beings in their "being for the sake of," the "*um zu . . ..*" In §15 Heidegger writes that "[a] thing is essentially something in order to [*Zeug ist wesenhaft etwas, um zu*] . . . ."[22] But how does this direct

17. *BT,* 61 (*SZ,* 65).

18. This is precisely what Heidegger is attempting in Part II, §70, "The Temporality of the Spatiality Characteristic of Dasein." *BT,* 336 (*SZ,* 367).

19. *BT,* 57–83 (*SZ,* 63–89).

20. *BT,* 65 (*SZ,* 69).

21. *BT,* 62 (*SZ,* 66).

22. *BT,* 64 (*SZ,* 68).

attention to an engagement of beings that is not in terms of entities present at hand? "Being for the sake of" occurs out of dasein's openness in its being always concernfully engaged with the being of beings. It occurs, in other words, by virtue of dasein's only way of being, i.e., as shown in Chapter 2, by being in the open with beings. "With its facticity, the being-in-the-world of Dasein is already dispersed in definite ways of being-in. . . . These ways of being-in have the kind of being of taking care."[23] Heidegger calls this concernfull dasein (being-t/here) "taking care" (*Besorgen*).[24] Out of this taking care entities show themselves in their "being for the sake of" (*um zu . . .*), and out of this "being for the sake of" spatialities are disclosed and entities find their place.[25] It is in light of this same play of dasein's being in the open with beings that the regions traditionally taken to be the vessel-like spaces that contains entities are found.

Heidegger argues that a "there" always belongs to place. "Place is always a determinate over there [*Dort*] or there [*Da*] belonging to a thing."[26] This "there" refers to dasein's taking care in that this mode of being occurs out of the entity's handiness, or the entity's being for the sake of (*um zu . . .*). The structure of taking care (*Besorgen*), out of which the entity is disclosed in its being for the sake of (*um zu . . .*), also orients the "there." The "there" arises out of the being for the sake of this or that. In the same way, one may go further and say that the network of places one calls a region is only discoverable in light of dasein's taking-care and out of the "for the sake of" of each event of being. As a network of particularities, the "for the sake of" of each entity orients spatiality in its "where to" (*Wohin*), or directionality. Heidegger identifies this "where to" with a kind of underlying "region" (*Gegend*); and this region is the possibility of place. The discussion gives a structural outline of how entities occur in their spatiality and find their unique places. What is crucial in this analysis is that Heidegger is considering nearness, place, and region not in terms of entities at hand, or ideal space. Rather, this analysis engages the temporality and finitude of the spatiality of events of beings in their disclosedness with dasein, or in light of dasein's concern-

23. *BT*, 52 (*SZ*, 56).

24. The term *Besorgen* in the analysis of dasein's being-in-the-world refers to the analysis of dasein's temporality that follows it, and specifically to dasein's fundamental attunement with beings through care (*Sorge*) in being-toward-death. Being-toward-death is discussed in Chapter 6.

25. *BT*, 96 (*SZ*, 103–4).

26. *BT*, 95 (*SZ*, 102).

full engagement with beings, i.e., in being in the open with beings. Thus, "region" (*Gegend*) does not point to any "thing," or to any transcendental idea, and spatiality is a matter of nothing other than dasein's engaged taking care in being in the open with beings. In short, the "*um*" of the *Umwelt* occurs not as objective or ideal spatiality, as a measurable spatiality of locations, but in the nearness of dasein and beings in their temporal disclosedness or finitude.[27] In the single turn in language from the "*um*" to the more originary sense of "*um zu . . .* ," in the move from understanding the "*um*" of the *Umwelt* in terms of entities at hand or ideal space to tracing it to its temporality and finitude, Heidegger opens a more originary aspect of spatiality. From this close look at §22 one can now see how his transformative appropriation of language and the tradition also occurs in the other two sections on spatiality, §23 and §24, where the transfiguration of spatiality is expanded and developed.

In §23, "The Spatiality of Being in-the-World," Heidegger discusses the particular way of dasein's being-in. There, Heidegger shows that dasein's being-in occurs spatially as "de-distancing" (*Ent-fernung*) and "directionality" (*Ausrichtung*).[28] These two terms require further discussion and clarification. Heidegger writes that "Dasein is essentially de-distancing [*Ent-fernung*]."[29] *Entfernung* generally means "separating" or "taking apart." But Heidegger writes it with a hyphen between "*ent*" and "*fernung,*" which means "distance," in order to point to the fundamental character of dasein's spatiality in taking-care as de-distancing. *Ent-fernung* does not mean overcoming a measurable "distance" or "space";[30] the prefix "*ent*" in German has the sense of "letting free." *Ent-fernung* refers to the letting free of spatiality that occurs in dasein's being in the open with beings. At the same time, *Ent-fernung* is an existential, constitutive part of dasein's being. In other words, de-distancing is the bringing near of beings in their nearness, in their disclosedness or appearing. Heidegger's term indicates that before traversing spaces in order to reach, see, hear, and so forth, one must have already brought beings into nearness by "taking care." One might relate this insight to the advent of modern

27. On "nearness" (*Nähe*) in Heidegger's discussion of spatiality in *Being and Time,* see Emil Kettering, *Nähe: Das Denken Martin Heideggers* (Pfüllingen: Neske, 1987).

28. "Die Räumlichkeit des Daseins wird characterisiert durch 'Ent-fernung' und 'Ausrichtung.'" Kettering, *Nähe: Das Denken Martin Heideggers,* 109.

29. *BT,* 97 (*SZ,* 105).

30. As Heidegger points out in §23, "Dasein does not traverse, like an objectively present corporeal thing, a stretch of space, it does not 'eat up kilometers'; nearing and de-distancing are always a heedful being toward what is approached and de-distanced." *BT,* 98 (*SZ,* 106).

physics in the face of the animism of Aristotelian metaphysics. Heidegger's discussion points to a level of engagement in being-in-the-world that has already occurred in order to arrive at Aristotle's animism, and that leaves the modern physical world of action at a distance by pointing to the fundamental engagement in that which has always already taken place before any interpretation of spatiality can be raised.[31]

In the same section Heidegger adds that de-distancing has the characteristic of *Ausrichtung*, directionality[32]: "As being-in which de-distances, Dasein has at the same time the character of directionality."[33] Again Heidegger engages directionality in terms of dasein, and shows that directionality is only possible in light of a fundamental engagement in being-in-the-world that is only and necessarily presupposed by directionality. In this passage on directionality, Heidegger refers critically to Kant's discussion of directionality in "*Was heisst: sich im Denken orienteren.*" Heidegger points out that directionality cannot be said to be in the body as *subjectum*, nor to have "some sense" of direction, unless one is willing to go back and fundamentally reexamine this ambiguous sense of directionality in terms of being-in-the-world.[34] In short, in this section, Heidegger translates both "distance" and "directionality" by engaging them in terms of dasein's taking-care, a mode of being that occurs out of dasein's concrete finitude and temporality as ultimately figured in the structure of dasein's care (*Sorge*).[35]

The last section of Heidegger's discussion of spatiality, "The Spatiality of Dasein and Space" (§24), concerns "space" in its thematic manifoldness. According to Heidegger, dasein's spatiality has the character of letting beings be in their spatiality. In its being-in-the-world, dasein has always already encountered a world.[36] This occurs in disclosedness, a "freeing of beings for a totality of relevances."[37] The spatiality of being-in, de-distancing, and its directionality, are the ways of spatiality that are

31. The question of the relationship between Heidegger's phenomenological insight and contemporary physics must remain open here, since in the work of Heisenberg and in quantum mechanics the phenomenon comes to be viewed by scientists as subject to engagements that configure, if not precede, perceptible phenomena (entities), their measurements, and calculations. In the specific case of Heisenberg I should point out that although he has a clear sense of such engagements, he understands them in terms of psychology, an approach that Heidegger certainly rejects as already grounded in a metaphysics of subjectivity.

32. "Als ent-fernendes In-der-Welt-sein hat Dasein zugleich den Character der 'Ausrichtung.'" Kettering, *Nähe: Das Denken Martin Heideggers*, 111.

33. *BT*, 100 (*SZ*, 108).

34. *BT*, 103 (*SZ*, 110–11).

35. See n. 24; and Chapter 6.

36. *BT*, 102 (*SZ*, 110).

37. Ibid.

constitutive of dasein. Thus, Heidegger states that "the freeing of a totality of relevances is equiprimordially a letting something be relevant in a region which de-distances and gives direction. It is a freeing of the spatial belongingness of things at hand."[38] In other words, dasein's spatiality is nothing other than the being in the disclosedness and possibility of the spatiality of inner-worldly beings with dasein (being-t/here).[39] This does not point to a spatiality in the sense of the "space" of entities at hand, but to the disclosive character of spatiality, since spatiality belongs to dasein's being-in-the-world as a constitutive de-distancing and orienting, elements that occur in the engaged taking care of dasein's facticity and that are the preconceptual grounds for the taking place of objective and ideal entities. Heidegger writes that "[s]pace is neither in the subject nor is the world in space. Rather, space is 'in' the world since being-in-the-world for Dasein has disclosed space."[40]

Following his discussion of dasein's spatiality, Heidegger considers the spatiality given in calculation and measurement. He also views as part of this study of spatiality the work of traditional phenomenologists such as Oskar Becker. According to Heidegger, both empirical studies of spatiality and the deduction of transcendental spatiality are relations already grounded in dasein's being-in-the-world and the fundamental spatiality of this mode of being.[41] In short, in §24, as in the previous two sections, Heidegger reintroduces traditional terms of spatiality by thinking them out of dasein's (being-t/here's) ontico-ontological way of being.

Withdrawal

However powerful Heidegger's insight concerning spatiality, his transformative appropriation of language for the sake of the letting be of the spatiality of the disclosedness of events of beings (i.e., dasein's spatiality)

38. Ibid.

39. "Die Räumlichkeit des Zuhandenen gründet in der Räumlichkeit des Daseins als In-der-Welt-sein." Kettering, *Nähe: Das Denken Martin Heideggers*, 112.

40. *BT*, 103 (*SZ*, 111).

41. Heidegger writes that "[t]he 'formal intuition' of space discovers pure possibilities of spatial relations. Here there is a series of stages laying bare pure homogeneous space, going from the pure morphology of spatial shapes to *analysis situs* and finally to the purely metrical sciences of space. In this present study we shall not consider how these are all interconnected. In our problematic we wish solely to establish ontologically the phenomenal basis for the thematic discovery and working out of pure space." *BT*, 104 (*SZ*, 112).

occurs as a withdrawal from his own attempt.[42] As Edward Casey points out in *The Fate of Place,* Heidegger's discussion of dasein's spatiality reduces it to a question of instrumentality. "Heidegger proceeds to an analysis of place (and region) that has little to do with caring and cherishing and everything to do with instrumental values . . . neither sheer location in world-space nor dwelling in depth, but place-as-pragmatic—as the realm of work-on things."[43] Casey's critique indicates that Heidegger fails to thoroughly engage in a discussion of spatiality when he does not analyze the existentiality of dasein's being-in-the-world in its spatiality. As Casey puts it, in *Being and Time* Heidegger only comes to think the full spatiality of the occurrences or events of beings by "indirection."[44] Here Heidegger's discussion seems to withdraw from spatiality in order to engage the issue in terms of instrumentality. But how is one to understand this failure to fully engage the spatiality of being-in-the-world? Would a "spatial" analysis of the "existentials" of being-in (i.e., falling or throwness) offer a grounding for a discussion of the spatiality of beings?[45] The answer is no, since this would not prevent or address the withdrawal. As we will see, this withdrawal is forced by Heidegger's transcendental turn in *Being and Time.*

The withdrawal of Heidegger's logos points to the central problem with *Being and Time.* In it, Heidegger attempts to open the question of being that has been covered over by traditional ontology and its attachment to the logic of presence. He does this by giving an analysis of the disclosedness of events of beings in terms of a temporal horizon, or of dasein's being-in-the-world. By addressing the question of being in this way he sets up a transcendental structure of being, wherein dasein must transcend itself in order to come to its temporality, to the disclosedness of being/time as such. As a consequence of this structure, spatiality slips beyond Heidegger's articulation, between a horizon of transcendental temporality and dasein's ontic presence. The duality that this articulation creates, this way of thinking the ontological difference between being and beings, leaves Heidegger no room for conceiving facticity and its spatiality otherwise

42. As well as in his earlier insight and engagement of the spatiality of events of beings in §12, Heidegger points to "being by" (*sein-bei*) and "inhabiting" (*wohnen*), and to "taking care" (*besorgen*) as the ways of the spatiality of dasein. See Hubert Dreyfus, *Being-in-the-World* (Cambridge: MIT Press, 1991).

43. Casey, *The Fate of Place,* 246.

44. Casey, *The Fate of Place,* chap. 11, "Proceeding to Place by Indirection: Heidegger," 243.

45. Ibid.

than in terms of either the necessities of the logic of ontic (objective) presence, or pure temporality. As a result, in §§22–24 Heidegger must engage spatiality in terms of the necessities and the logic of objective presence. This is clear when one briefly looks back at these sections.

In §22 Heidegger discusses spatiality in terms of the necessary "region" (*Gegend*) that underlies places and their diverse networks. Then, in §23, he discusses being-in in terms of the possibility of letting appear of beings out of dasein's characteristic de-distancing (*Ent-fernung*), and directionality (*Ausrichtung*). Finally, in §24, Heidegger speaks of dasein's spatiality only in terms of the letting be of the spatiality of entities in their coming to presence, as a "giving space" in the sense of "arranging" (*Einräumen*),[46] and he organizes spatiality into a diversity of thematic treatments according to various modes of circumspection that let the spatiality of beings come to presence.[47] Although in his discussion of spatiality Heidegger attempts to engage presencing in terms of essential temporality and not in terms of the presence of entities at hand, the discussion remains concerned with the presence of beings, and therefore does not take up the phenomena in the full sense of presencing, i.e., by engaging the withdrawal that is always operative in the coming to presence of occurrences of beings and of thought. In short, Heidegger's discussion of spatiality in §§22–24 of *Being and Time,* rather than engaging spatiality, withdraws from the very withdrawal and absencing operative in occurrences or events of beings and of thought.

Enactments

Heidegger's discussion reinscribes the language of "space" into a discourse of presence, and in doing so fails to engage the withdrawing and absencing, which, according to him, is fundamental to the presencing of occurrences of beings, and in particular fundamental to the motion of his thought and language (as indicated in Chapter 2). How is one to understand Heidegger's philosophical discourse in light of this withdrawal, this failed leap? First of all, Heidegger's leap remains an ungrounded event.

46. *BT,* 103 (*SZ,* 111). *Einräumen* has the sense of ordering and organizing, the configuration of spaces and places; it is a word used in architecture.

47. *BT,* 104 (*SZ,* 112).

The language is ungrounded in that it has been severed from the tradition of objective and ideal presence (through his critique of metaphysics and transcendental philosophy in his engagement with Descartes' ontology). In spite of the orientation of Heidegger's discussion of spatiality toward presence, the tradition of objective and ideal presence cannot ground or give direction to his analysis. Furthermore, the events of language and thought are ungrounding, because in Heidegger's use of the tradition's terminology there is always an inherent opposition, resistance, and decentering of that tradition. Hence the very function of the traditional terms in Heidegger's discussion will always transform, or at least interrupt, any traditional claim to ever-present, unchanging origins, ideas, essences, or transcendental structures of possibility for the occurrences of beings. As such, Heidegger's discourse appears as an event that arises from and comes to pass on exilic grounds: the event of thought occurs in the absence of the possibility of finding meaning or origin in ever-present, unchanging ideas, and at the same time, it occurs in the enactment of a motion open to the essential transformative force of its events. Furthermore, Heidegger's language enacts a passage that remains beyond presence, and as such intimates its alterity. In the withdrawal from the absencing that is operative in presencing, Heidegger's thought conceals its event, the ungrounded and ungrounding character of the discussion in its taking place.[48] In view of these two performative aspects of Heidegger's discussion of spatiality, the alterity of thought and exilic grounds are certainly felt. Even if Heidegger's discourse withdraws to questions of presence, the motion of that thought indicates the full difficulty and evanescent passage of thought in its temporality and finitude .

Conclusion

Chapter 1, "Transgressions," initiates a journey of philosophical discourse to the limits of the world of objective and ideal presence as a figure that supplements and accompanies the difficulty of thought in engag-

48. I am not suggesting that concern can be separated from presence, but indicating how presence is always a difficulty that marks the possibilities of engaging alterity. This figures an ever-operative temptation and danger in philosophical discourses, a danger that is clearly overpowering to thought—for example, when the difficulties of thought are framed into ideologies and agendas of action, i.e., in the politicization of philosophy.

ing alterity. Such a journey puts the engagement of events of beings and of thought in their presencing and absencing in danger, because the logos, on its return, spreads like a vast net of the conceptual determinations of events of beings. In the case of Plato, Aristotle, and the tradition that follows them, this return of the logos produces a story about beings, about the entities objectively present at hand. The logos functions as a mimetic tool that represents all beings, and even itself. It is as if on its return the logos returns being to beings through its representation of them; and in those representations events of beings have sense. At the same time, by the same economy, only as such a representation does the logos have sense.

In the various discussions of spatiality in the previous chapters a series of traces mark a path of engagement with the alterity and exilic grounds of thought, from the difficulties figured by *chora* to the difficulties found with the motion of thought concerning spatiality as a figure of alterity in *Being and Time.* This path leads to the engagement of Heidegger's apophantic logos in its transformative and reappropriating passage. The previous discussions in this chapter and others have shown that Heidegger's thought comes to pass and is sustained by a leap that always places it outside a possible return to the traditional "meaning" of world and thought as events grounded on ever-present, unchanging origins, principle, and structures. At the same time, this leap is performed by a thought that in its critique appropriates and decenters this tradition. Furthermore, in this transformative motion, Heidegger's thought overcomes its event, its configuration or articulation of events of beings in transcendental terms as well as in its concern with presence. Lastly, this thought can only perform intimations of the matter to be thought; it is always a thought on the way. These insights indicate that this thought occurs as and remains an exilic event. This also means that it is not an event that can be traced back to or beyond its passage. It occurs on the exilic grounds enacted by the passage of thought. Thus, it can neither be traced to other beginnings nor can it be conceived teleologically. Here the logos will not "return," since it will not travel beyond its ephemeral events. Rather, it will perform a "re-turning," a transformative appropriation of the tradition in its event. Furthermore, this re-turning will not cover over the alterity of beings and of their events. The difficulties with Heidegger's engagement of the spatiality of events of beings in their disclosedness mark points where the apophantic logos engages and, to a certain extent, remains with the alterity of beings and of its event. Here, what would seem for the tradition a loss of all senses of being through the uncovering of the alterity and exilic grounds of thought, instead figures the beginning of an engagement with all sense of beings and thought in their presencing/absencing. In light of these conclusions just made, one cannot say that the sense of philosophical discourses has been lost in the transformative reappropriation of the tradition (*Destruktion*), but rather that in taking up Heidegger's thought a series of moments at the limit of all sense of beings and of thought have been found.

6

Exilic Passages

Dasein's Being-Toward-Death

Introduction

The last three chapters have traced the figure of the alterity and exilic character of Heidegger's thought by engaging specific moments in his discourse on the temporality of the disclosedness of the occurrences or events of beings. In these discussions spatiality appears as a figure of alterity and exilic grounds; it appears as an operative difficulty that continues to interrupt Heidegger's discourse. Throughout these chapters, the discussion of Heidegger's thought in *Being and Time* is bilateral: on the one hand, it concerns what Heidegger says, and follow closely the structural aspects of his analysis of dasein in light of his focus on temporality; on the other hand, these intentional aspects of Heidegger's discourse are also engaged in their performative character. The performative aspect of Heidegger's work reveals a series of interruptions, suspensions, slippages, and withdrawals operative in the very motion or passage of his thought. This chapter takes up the alterity and exilic grounds figured by these aspects of Heidegger's thought, and engages them in light of Heidegger's own discussion of dasein's finitude and temporality (dasein's being-toward-death) in *Being and Time.* This places the performative aspects of Heidegger's thought and the intimations of alterity and exilic grounds in the context of Heidegger's own project. In relating this discussion of alterity and exilic grounds to Heidegger's analysis of dasein's being-toward-death, two central insights appear. First, Heidegger's uncovering of dasein's being-toward-death leads toward a necessary rethinking of philosophical discourse in terms of the alterity and the exilic aspects of events of beings and thought as figured by dasein's temporality and the issue of dasein's spatiality. Second, in view of thought's alterity and exilic grounds, the thought of *Being and Time* can be engaged through a deconstruction of its attempt to analyze temporality as the origin of events of beings. In short, this chapter rethinks the previous insights drawn from the various discussions thus far in light of dasein's temporality; and by doing this, shows that Heidegger's own project leads toward the engagement of the alterity and exilic character of beings and thought—in this sense, also pointing toward the engagement of the very event of thought that is *Being and Time* in terms of its alterity and exilic grounds.

Being-Toward-Death: Dasein's Alterity

Heidegger's analysis of dasein's being-in-the-world in Part I of *Being and Time* leads to his analysis of the fundamental temporality of this structure in the book's second part, i.e., to the analysis of dasein's "ekstatical being"[1] or its "being-toward-death" (*Sein zum Tode*).[2] This path may be quickly situated within *Being and Time*'s general project. Herein, Heidegger attempts to trace the presencing of events of beings back to their ontological and temporal grounds by uncovering the events of presencing, both as events of disclosedness and in their temporal horizon—i.e., by thinking them out of dasein's being-in-the-world. According to Heidegger, being-in-the-world occurs as the event of and possibility for the disclosedness of all beings (*Lichtung*). This occurs out of dasein's "care"

1. "Die Phänomene des zu . . . , auf . . . , bei . . . offenbaren die Zeitlichkeit als das 'εκστατικον' schlechthin." *BT,* 302 (*SZ,* 329).
2. "Die Sorge ist Sein zum Tode," *BT,* 303 (*SZ,* 329).

(*Sorge*), out of dasein's being always already in the open with beings in an ontological engagement that gives the question of being its way of being. This care occurs out of dasein's finitude and temporality, i.e., dasein's essential "being-toward-death."[3] In other words, according to Heidegger, the ground for the structure of dasein's concernfull being-in-the-world is dasein's finitude and temporality, which occurs as or in dasein's (being-t/here's) being-toward-death. This is what Heidegger indicates when he states that "[t]he primordial unity of the structure of care lies in temporality."[3]

Temporality can initially be understood in terms of dasein's being-toward-death. Heidegger notes in §65 that "[c]are is being-towards-death." This means that the presencing of events of beings occurs in light of the finitude that is characteristic of dasein's way of being-in-the-world. "Care is being-towards-death. We defined anticipatory resoluteness as authentic being towards the possibility that we characterized as the absolute impossibility of dasein. In this being towards the end, dasein exists authentically and totally as the being that it can be when 'thrown into death.' It does not have an end where it just stops but *it exists finitely*."[5] This last sentence indicates dasein's way of being: unlike entities at hand, and unlike the ontic, the ontico-ontological being-in-the-world of dasein occurs always in light of its finitude. Just as dasein does not choose whether to be in the world concernfully with others or not (this is the only way dasein exists), its finitude always occurs in the open with beings and occasions their disclosedness. Here, finitude does not mean death, understood as the end of an entity's life or activity. Dasein's finitude indicates a way of being-in-the-world that has arisen in light of a certain sense of temporality. And dasein's finitude, Heidegger writes in the same section, "does not primarily mean a stooping, but is a characteristic of temporalizing itself."[6]

In *Being and Time* Heidegger interprets this temporalizing primarily in terms of dasein's (being-t/here's) moment and "futurity." According to Heidegger, dasein occurs most authentically as a "there" (*Da*), as a disclosive event constituted as a "moment" (*Augenblick*).[7] This moment has

3. Ibid.
4. *BT,* 301 (*SZ*, 327).
5. *BT,* 303, italics in original (*SZ*, 329, 330).
6. *BT,* 304 (*SZ*, 331).
7. "Im Modus der ursprünglichen Zeitlichkeit eingeschlossen bleibt in Zukunft und Gewesenheit. Entschlossen hat sich das Dasein gerade zurückgeholt aus dem Verfallen, um desto eigentlicher im 'Augenblick' auf die erschlossene Situation 'da' zu sein." *BT,* 302 (*SZ*, 328).

a threefold structure: "future" (*Zukunft*), "having been" (*Gewesenheit*), and "present" (*Gegenwart*).[8] The unity of this threefold structure is indicated phenomenologically by dasein's modes of everydayness as "toward itself" (*Auf-sich-zu*), "back to" (*Zurück auf*), and "letting something be present" (*Begegnenlassens von*).

> Future, having been, and present show the phenomenal characteristics of "toward itself," "back to," "letting something be encountered." The phenomena of toward . . . , to . . . , together with . . . , reveal temporality as the εκστατικον [*ekstatikon*] *par excellence. Temporality is the primordial "outside of itself" in and for itself.* Thus we call the phenomena of future, having been, and present, the *ekstases* of temporality. Temporality is not, prior to this, a being that first emerges from *itself*; its essence is temporalizing in the unity of the *ekstases.*[9]

The moments of the temporality of everydayness arise in light of the *Augenblick,* dasein's ekstatical temporality. This "moment" figures an event that always occurs beyond presence, in that it is essentially constituted in a single mode of passage, in light of futurity and memorial aspects.[10] This being-toward-itself of dasein does not belong to the phenomena of future, past, and present as understood in the common sense of "time," nor does temporality emerge "from itself" prior to the *ekstases* of the *Augenblick.* Thus, dasein's care is neither transcendentally oriented by a pure temporality prior to the event of the moment, nor is it reducible to ontic temporality, any more than spatiality is reducible to ontic "spaces."[11]

In *Being and Time,* Heidegger understands the moment in terms of dasein's finitude, in terms of dasein's being-toward-death, which under-

8. Ibid.

9. *BT,* 302, italics in original (*SZ,* 329).

10. I refer here not to the past of a historical consciousness, but to the element of involuntary memory that belongs to the moment. See Charles Scott, *The Time of Memory* (Albany: SUNY Press, 1999).

11. "Was heißt das, daß ich so etwas wie Raum akzeptiere? Ich akzeptiere, daß es so etwas wie Raum gibt, und mehr als das, daß ich eine Beziehung zum Raum und zur Zeit habe. . . . Über dieses Akzeptierte kann nicht mehr der Physiker etws sagen, sondern nur noch der Philosoph." Heidegger, *Zollikoner Seminare* (Frankfurt am Main: V. Klostermann, 1987), 35; Cf. *SZ,* §70, 367: "If in the course of our existential interpretation we were to talk about the 'spatio-temporal' determination of dasein, we could not mean that this being is objectively present 'in space and also in time.'"

lies or sustains dasein's being-in-the-world, taking care (*Besorge*), and care (*Sorge*). Heidegger writes that "the primary phenomenon of primordial and authentic temporality is the future."[12] The future, the being-toward that characterizes dasein authentically, neither refers to a projection toward a potential actuality nor does it hold to a teleological motion.[13] Rather, it indicates a disclosedness that occurs out of "the impossibility of Dasein" (*der schlechthinnigen unmöglichkeit des Daseins*). This impossibility does not represent the end of a finite entity among other entities (the human subject); dasein's "thrownness" (*Geworfenheit*) has the sense of a certain "futurity," a projection that keeps dasein in place in light of a "not-yet," a yet to come. "Dasein always already exists in such way that its not-yet [*Noch-nicht*] belongs to it."[14] Dasein's not-yet indicates a certain absence operative in all dasein (being-t/here). This takes us further into the issue of temporalizing dasein by indicating that both dasein's being-in the-world and the disclosedness of events of beings within this horizon occur out of a certain absencing or withdrawing that belongs to dasein's event in its futurity. This is what Heidegger means when he writes that dasein's temporality is "the primordial 'outside of itself.'" This statement can be interpreted as indicating that dasein is an entity that must go beyond itself in order to reach its own temporality or futurity. However, Heidegger's thought goes beyond such a transcendental interpretation. In speaking of a certain "primordial 'outside of itself,'" Heidegger's words indicate an otherness in the event of dasein (being-t/here). In uncovering dasein's modality of being-toward-death Heidegger brings forth that aspect of dasein that will always remain beyond objective and ideal presence (hence also representational thought). And yet this futurity is not nothing, but it is constitutive of the events of disclosedness of dasein and of beings. The operative aspect of dasein's events (futurity) that will remain beyond presence suggests that dasein's configuration and presencing will always occur beyond presence, in a certain absence and alterity.[15] Therefore, according

12. "Das primäre Phänomen der ursprünglichen und eigentlichen Zeitlichkeit ist die Zukunft." *BT,* 303 (*SZ,* 329).

13. As would be the case in either an Aristotelian teleology, or in Hegel's historical motion toward absolute knowledge.

14. *BT,* 226 (*SZ,* 243); see also *BT,* 224–28 (*SZ,* 242–46), and *BT,* 136 (*SZ,* 145) on the relationship between the not-yet and the possibility of understanding.

15. In *Language and Death,* Giorgio Agamben gives a beautiful indication of dasein's alterity. "If this is true, if being its own Da (its own there) is what characterizes dasein (Being-there), this signifies that precisely at the point where the possibility of being Da, of being at home in one's

to Heidegger's analysis, the disclosedness of events of beings and thought in their presencing will only occur in light of a certain otherness to presence. The word *ekstatikon* echoes dasein's alterity, this absence in presencing, because it literally means "being in departure" or "being displaced," and in this way, being always other and beyond—to a certain extent, being foreign.[16] Even if one were to interpret dasein in terms of the transcendental articulation of *Being and Time,* as the presence of an entity that must then transcend itself toward a horizon of temporality, dasein's futurity puts this modality of being beyond such objective presence. Dasein occurs in presencing, i.e., in coming to pass in the enactment of a certain futurity. In Heidegger's words, dasein does not occur as "a summative together that is outstanding, but rather a not-yet that Dasein always has to be."[17]

The Alterity of Thought

The alterity of dasein points to the alterity of the occurrences or events of beings, and particularly to the alterity of thought, in their presencing disclosedness. This follows directly from Heidegger's thematic treatment of the presencing or disclosedness of events of beings in terms of their temporal horizon, or dasein (being-t/here).[18] In light of this project, one can see that the absencing of being enacted by dasein's (being-t/here's) futurity or not-yet marks the occasion for the presencing of beings. This insight also applies to thought. Indeed, Heidegger first mentions dasein's futurity in *Being and Time* when he discusses understanding (*Verstehen*). According to Heidegger, "[*u*]*nderstanding is the existential being of the*

own place, is actualized through the expression of death, in the authentic mode, the Da is finally revealed as the source from which a radical and threatening negativity emerges." *Language and Death: The Place of Negativity,* tr. Pinkis & Hardt (Minneapolis: University of Minnesota Press, 1991). Agamben's project moves toward engaging "the outer limits of Heidegger's thought," and in this sense it is analogous to the present work. One single difference is that Agamben seems to be concerned with "negativity" and "ground," whereas this study engages the outer limit of Heidegger's thought in terms of finitude or ephemeral passages and the transformative character of negativity or absence. This difference appears to me not as a criticism of Agamben's work but as another possible engagement of Heidegger's thought at its limit.

16. The Greek word *ekstatikon* also recalls *ekstasis,* which can mean astonishment.

17. *BT,* 227 (*SZ,* 244).

18. *BT,* 1 (*SZ,* 1).

ownmost potentiality of being of Dasein in such a way that this being discloses in itself what its very being is about."[19] But what is it that is disclosed in dasein? Heidegger goes on to discuss understanding and explains that it occurs as "projecting" (*Entwerfen*): "understanding in itself has the existential structure which we call project." This projection occurs in the enactment of dasein's not-yet. First of all, understanding occurs as projecting, and therefore in a structure analogous to dasein's futurity or not-yet. "Projecting has nothing to do with being related to a plan thought out, according to which Dasein arranges its being, but, as Dasein, it has always already projected itself, and is, as long as is, projecting." But this projecting is nothing other than dasein's factical way of being-in-the-world.

> Because of the kind of being which is constituted by the existential of projecting, Dasein is constantly "more" than it actually is, if one wanted to and one could register it as something objectively present in its content of being. But it is never more than it factically is because its potentiality of being belongs essentially to its facticity. But, as being possible, Dasein is also never less. It is existentially that which it is *not yet* in its potentiality of being. And only because the being of the there gets its constitution through understanding and its character of project, only because it is what it becomes or does not become, can it say understandingly to itself: "become what you are."[20]

In other words, understanding occurs as the enactment of dasein's alterity in its factical aspect or concrete finitude and temporality. Understanding occurs only in alterity, and as such enacts the disclosedness of events of beings and thought. At the same time, this sense of understanding remains beyond objective and ideal presence and representational thought. Therefore, in the uncovering of the alterity of thought as figured in understanding, one also finds the difficulty of engaging events of thought in their alterity (understood as events that remain beyond presence and representation).

The passage above places understanding in dasein's concrete finitude. This aspect of thought and of the presencing of beings is fully engaged by

19. *BT,* 135 (*SZ,* 144).
20. *BT,* 136, italics in original (*SZ,* 145).

Heidegger in a later version of what was to be the third part of *Being and Time,* in his reading of Kant's *Critique of Pure Reason.*[21] In contrast to other thinkers' "deontologizing" or "analytical" readings of Kant's *Critique,* Heidegger does not finds in Kant's thought a system of a priori concepts and synthetic functions of reason that gives form and remains a structure of possibility for and above the "world." In his reading Heidegger engages Kant's thought as "a finite metaphysics."[22] According to Heidegger, in the A Edition of the *First Critique,* Kant touches on the disclosedness of beings in its occurrence, or "possibility" in the grounding of knowledge not on the transcendental logic (as is the case in the B edition) but on the transcendental power of imagination as the synthesis of intuitions and concepts.[23] This possibility of knowledge and experience is given in the finite character of imagination, which means that possibility is ultimately understood not transcendentally but in light of the finitude of imagination. The finite imagination, the passing glance or moment (*Augenblick*) of mortality, occurs as the possibility for the occurrences or events of beings. This finitude is also operative in all thought and its conceptual determinations.[24] This insight sustains Heidegger's claim in *Being and Time* that dasein's fundamental ontological structure is held together by the temporality of dasein's finitude or being-toward-death.[25] If there is any possibility for thought and for the engagement of events of beings, it is in the event of dasein's concrete finitude.

In light of the alterity of thought and its concrete finitude the task appears of engaging thought in such manner that these aspects will not be covered over; this means understanding, withstanding, standing under thought's alterity in its concrete passages. At this point philosophical discourse appears as a matter of thinking in the awareness and enactment of

21. Heidegger, *Kant and the Problem of Metaphysics,* tr. Richard Taft (Indianapolis: Indiana University Press, 1990).

22. Emmanuel Kant, *Critique of Pure Reason,* tr. Norman Kemp Smith. (New York: St. Martin's Press, 1929); *Kritik der reinen Vernunft* (Hamburg: Felix Meiner, 1956). See Hans-Georg Gadamer, "Kant and the Hermeneutical Turn," in *Heidegger's Way,* tr. John W. Stanley. (Albany: SUNY Press, 1994): 57.

23. Heidegger, *Kant and the Problem of Metaphysics,* Part II–III, tr. Richard Taft (Indianapolis: Indiana University Press, 1997); *Kant und das Problem der Metaphysik, GA* 3. Heidegger is emphatic about the role of imagination in light of "possibility."

24. *Kant and the Problem of Metaphysics,* Part IV. Cf. Heidegger, *Phenomenological Interpretation of Kant's Critique of Pure Reason,* tr. Parvis Emad and Kenneth Maly. (Indianapolis: Indiana University Press, 1997); *Phänomenologische Interpretation von Kants Kritik der reinen Vernunft, GA* 25, §§25–26.

25. *Kant and the Problem of Metaphysics,* Part IV: b.

its alterity. Perhaps here one may speak of a welcoming of the other and of the foreign, in the need for the enactment of thought in light of its alterity. This is as much an issue of the nearness or the intimacy of thought to its events as it is one of thought's foreign character to its very passages, since, in touching on the alterity of thought, one will have always found what is ownmost to thought, and with it, what must remain foreign.[26] We are speaking of an engagement with thought's alterity that goes beyond representational and conceptual determinations of thought, in the enactment of a passage that, rather than presenting, will recall the loss essential to thought and to the presencing of events of beings.[27]

Excursus

Before taking up the alterity of thought,[28] it will be helpful to indicate how the figure of spatiality can be engaged in light of the futurity of the disclosedness of events of beings and thought. Or, to put it in question

26. Although this is not the place to even begin a discussion of this topic, I want at least to point out that I think it is this engagement of thought in its alterity that marks the "place" where Heidegger's thought encounters the thought of "the Greeks." See Heidegger "Vom Wesen der Wahrheit," *Wegmarken, GA* 9; "Aletheia (Heraklit Fragment 16)," in *Vorträge und Aufsätze*; *Vorträge und Aufsätze* (Pfüllingen: Neske, 1985); and "*Aletheia* (Heraclitus, Fragment B16)," in *Early Greek Thinking,* tr. David Farrell Krell and Frank A. Capuzzi (New York: Harper & Row, 1975). Cf. Heidegger, *Heraklit, "Der Anfand des Abendländichen Denkens (Heraklit),"* *GA* 55; "Der Spruch des Anaximander," in *Holzwege, GA* 5; "The Anaximander Fragment," in *Early Greek Thinking.*

27. I take this to be the task taken up by Heidegger in *Contributions to Philosophy,* first published in 1936, his second major work. It is the task that is also engaged by Giorgio Agamben's work on voice and text, and already begun in *Language and Death* (Minneapolis: University of Minnesota Press, 1991) and *Infanzia e Storia* (Turin: Einaudi, 1978); as well as by Maurice Blanchot in *The Writing of the Disaster* (Lincoln: University of Nebraska Press, 1995); and by Jacques Derrida's work on the margins of literature and philosophy (for example, *Margins of Philosophy* [Paris: Minuit, 1972] and *Dissemination* [Chicago: University of Chicago Press, 1981]).

28. The section title "Excursus" indicates a certain resistance that seems to keep temporality and spatiality separate in discourse. Once again, as is the case with *Being and Time,* in the engagement of temporality, spatiality seems to have drifted from the discussion. What is the character of this resistance in discourse? How do spatiality and temporality figure in the event of philosophical discourse? Might the difference between text and voice indicate something of this operative tension between spatiality and temporality? For now, one can at least be aware that in its almost unnoticed reintroduction, spatiality seems to mark a deviation, an interruption in the discussion.

form, how does dasein's futurity touch on the issue of the spatiality of the disclosedness of events of beings and thought?

The futurity of dasein's being-toward-death leads beyond dasein's spatiality as formally presented in *Being and Time.* In light of dasein's "ekstatical being," its being-toward-death, Heidegger's statement that spatiality must be thought out of dasein's temporality calls for thinking spatiality not only in terms of presencing, but also and fundamentally in its temporality—and this means in the absencing figured by dasein's not-yet. As a result of dasein's ekstatical being the issue of the "ekstatical spatiality" of the disclosedness of events of beings arises.[29] This points beyond Heidegger's discussion of spatiality in *Being and Time,* because the issue of spatiality points now beyond the presence of beings and toward the nonpresence operative in all presencing events. Ekstatical spatiality refers thought to a spatiality that cannot be grasped in terms of objective and ideal presence, since it does not occur as presence alone. In this respect the "*ek*" of dasein as the *ekstatikon par excellence* already directs us beyond presence. The root meaning of Heidegger's term for dasein indicates this being beyond presence. The root of the word *ekstases* is *ekstasis,* which means literally "standing outside." Furthermore, in this instance "*ek*" can only be understood in light of the alterity and concrete finitude of the disclosedness of thought and of events of beings. One must take care here not to fall back to a metaphysical or transcendental interpretation of "*ek.*" *Ekstasis* does not mean being outside some other being. The "*ek*" here marks the temporalizing character of dasein, and brings to spatiality the concrete finitude of events of beings and thought. When one looks back to the early passage in *Being and*

29. In addition, the withdrawing ground of dasein's being-in-the-world and the absence of an analysis of the spatiality of dasein's existential structure of care leave the issue of dasein's spatiality open. As shown, in §12 Heidegger makes an opening toward spatiality by differentiating between ontic "space" and dasein's own being-in; but then, in his discussion of the spatiality of dasein in §§22–24, he quickly withdraws from this opening. If one takes into consideration how the spatiality of being-in is never discussed as such, and that in fact Heidegger seems to evade it, then it becomes apparent that this evasion leaves the issue of spatiality open to extreme implications. These implications run to the very event of the withdrawal of the truth of being in its disclosedness as intimated in *Being and Time.* The issue of "intimation" is important here because, on the one hand, being is not thought in its truth in *Being and Time,* and because, on the other hand, the withdrawing character of the truth of being finds itself indicated by dasein's fundamental withdrawing character in its being-toward-death. Intimation then serves to remind us of the limit of the thought and language of *Being and Time* with respect to the truth of being. Furthermore, it indicates that there is already a certain articulation of the truth of being coming forth in *Being and Time.*

Time in which Heidegger introduces the ontological difference by differentiating between the objective and ideal "space" of entities present at hand and dasein's being-in, one finds the issue of concrete spatiality already announced.[30] In this passage Heidegger refers to dasein's spatiality in most concrete terms as "inhabiting" (*habito*), "dwelling" (*sein bei*), and "cherishing" (*colo*), all terms that must now be considered in light of dasein's futurity, but without abandoning the concrete finitude of disclosive events.

The Alterity of Thought: Thinking on Exilic Grounds

The recalling of thought to alterity leads to the concrete finitude of philosophical discourse; simultaneously, this concreteness unfolds on certain exilic grounds. When Heidegger encounters the disclosedness of events of beings and thought—not in an unchanging, ever-present, transcendental consciousness or operation of reason, but in mortal imagination—he recalls thought to its finitude, to its concrete and ephemeral events. *Dasein* (being-t/here), *Augenblick* (moment/glance), *Lichtung* (clearing): these terms figure the concrete finitude and ephemeral passage of events of beings and thought in their presencing. Here the term "concreteness" requires a brief clarification. What it indicates becomes clear when one considers that, for Heidegger, the horizon of temporality of all disclosedness of beings and thought is dasein. This being-t/here, this modality of being, does not refer to a subject or to an entity among others, nor to a historical or pragmatic fact. Dasein figures the modality of temporality already discussed as being-in-the-world, care, being-toward-death, and the not-yet. What is figured by concreteness, then, is the passage of beings and thought in their ephemeral eventuations. In other words, beings and thought are to be engaged in their temporality, and this means in their finitude.

For Heidegger the "finitude" of thought can only be engaged out of the temporality of its events, i.e., in light of the futurity figured by the structure of the *Augenblick*. This means that thought will only be engaged in light of dasein's alterity, the not-yet—in thought's engagement of its events in their alterity as figured by or enacted in their finite

30. *BT*, 50 (*SZ*, 54).

ephemeral passages. This puts thought beyond objective and ideal presence, and beyond representation. First of all, this is because in their futurity events of thought will remain always beyond their conceptual determinations, i.e., beyond self-certainty and self-determination, since their event will have always already gone beyond their conceptual determination. This realization begins to engage the alterity of thought. Secondly, with this indication of alterity appears the impossibility of thought's claims to absolute knowledge and ever-present, unchanging ideas, principles, structures, and teleologies, since thought cannot bring its events or the disclosedness of beings to presence, and therefore must remain outside of such objective and ideal interpretations of beings and thought. This means that the assertive representational power of philosophical discourse meets a limit, and that because of this limit, thought encounters an intimation of its exilic character, since it cannot refer to unchanging origins or principles in order to begin to understand its events.

Dasein's structure of temporality, and more specifically, its not-yet, opens thought to its exilic aspects. On the one hand, the nonpresence of events of disclosedness as figured in the futurity of dasein disrupts all possibilities of the return of events of thought to ever-present and unchanging original ideas or principles. On the other hand, this same futurity figures a certain open disposition in the disclosedness of events of beings and of thought in dasein, since these events do not occur as actualizing gestures of a potential that is already determined through a teleology, but enact events that remain toward a not-yet, and therefore toward an open horizon of disclosedness. We are speaking of a horizon that, unlike Heidegger's transcendental articulation of dasein in *Being and Time,* will not stand ahead of events of beings and thought, but will rather arise in the performative passages of beings and thoughts. In terms of dasein's futurity, thought will always come to pass in events that take it beyond its conceptual determinations in at least two ways: in the arising of thought in light of a projecting in a not-yet or absence—hence always occurring beyond presence—and in thought enacting a limit that will always already be open to events beyond its operative determinations. In short, the exilic character of thought is figured by the impossibility of thought's return to ever-present, unchanging ideas or origins, and in the transformative motion enacted by its events.

Figures and Enactments: Of Alterity and Exilic Thought in *Being and Time*

The previous chapters traced the alterity and exilic grounds of thought as figured by the difficult issues of spatiality. In these chapters spatiality first appears as a figure of the alterity and exilic grounds of beings and thought in the tradition (in the work of Plato, then of Aristotle), and then in Heidegger's engagement of events of beings and thought in terms of originary temporality in *Being and Time.* In light of the present discussion of dasein's being-toward-death and the alterity and exilic grounds figured by this sense of temporality and finitude, the difficulties figured by spatiality in these chapters serve as clear indications of the alterity and exilic character of thought as figured in *Being and Time.* It is not that Heidegger's aim is to give an account of ekstatical spatiality—as does, for example, Timaeus in Plato's dialogue, whose account requires three "kinds": being, becoming, and the strange third kind, *chora.*[31] Instead, Heidegger's treatment of spatiality is far from being a discussion of a "kind," a return to the logic of objective and ideal presence. In the motion of Heidegger's thought the issue of spatiality arises, and with it, one begins to encounter indications of the alterity and exilic character of his thought as enacted by his discourse on the originary temporality of events of beings and thought. The issues of alterity and exilic thought are figured by the moments of suspension, interruption, and indeterminacy that punctuate the discussions of spatiality. They are found in light of the twisting free of spatiality from traditional objective and ideal interpretations, as well as in the slipping of spatiality from Heidegger's transcendental articulation. They are also felt in the unfinished character of Heidegger's transformative reappropriation of the language of objective and ideal "space," in his attempt to begin to engage the spatiality of the disclosedness of events of beings. Finally, the exilic power of thought is engaged by following the transformative motion of Heidegger's thought in *Being and Time,* a thought that in its passage overcomes its very event as it leads beyond its transcendental discourse and toward another way of engaging events of beings. In these nonrepresentable moments thought echoes in its alterity and exilic passages.

31. Plato, *Timaeus,* 48e–49.

Seen in light of these issues, *Being and Time* figures the traces of an opening, the beginning of an engagement with the alterity and exilic grounds of philosophical thought. Furthermore, touching on thought's alterity and exilic character recalls all conceptual determinations to these grounds. At the same time, this insight serves as a beginning step toward the deconstruction or transformative reappropriation of Heidegger's thought in *Being and Time.* This reappropriation is figured in the previous chapters by the moments when the discussion moves beyond Heidegger's intended discourse on the originary temporality of the occurrences of beings—moments that are drawn out of the engagement of the performative aspect of Heidegger's thought. Such engagement shows again and again that Heidegger's analysis of dasein's temporality enacts a different and decentering motion in its passage, since throughout the discourse the idea of "origin" is decentered and interrupted, transformed, and forced to remain in indeterminacy by the operative roles of spatiality. Furthermore, these decentering events lead ultimately to the abandonment of the discourse, an abandonment brought forth by the very passage of Heidegger's thought in *Being and Time.* The deconstruction of Heidegger's discourse does not conclude with the exclusion or abandonment of temporality. Rather, as shown in this chapter, dasein's temporality figures the alterity and exilic character of thought. And yet, in light of the alterity and exilic character of the disclosedness of beings and thought, temporality can no longer operate as the "origin" for the occurrences or events of being, since it can be taken as an origin neither in terms of traditional discourses of objective and ideal presence nor in terms of Heidegger's transcendental articulation of the disclosedness of events of beings and thought.

Conclusion

It is now apparent that in view of dasein's futurity or not-yet, Heidegger's analysis leads to a thinking in alterity and out of exilic grounds. At the same time, recalling Heidegger's thought to such grounds also opens *Being and Time* to a deconstructive reappropriation. In light of thought's alterity and transformative motion Heidegger's discussion of temporality in *Being and Time* appears as the formal indication of that thought's self-overcoming event or passage. Heidegger engages dasein's temporality and ekstatical being, its not-yet, as the originary temporality of events of beings and thought, and this move figures the overcoming or transformative aspect of its very occurrence.

This chapter closes the second section and its engagement with Heidegger's thought in *Being and Time.* The following final chapter discusses Heidegger's later thought with respect to spatiality, alterity, and exilic thought. This series of encounters with Heidegger's thought begins with the introduction to this book, and does so by interruption: Most powerfully, the voice of Julia Kristeva first brings the shadow of alterity to *Being and Time.*[32] Kristeva points out that in order for alterity to be heard, a certain fall, a certain giving way of ground or self-certainty is necessary. This notion offers a way to understand what the present work accomplishes. With this chapter culminates an engagement with Heidegger's thought in *Being and Time* that marks that thought's "failing" in its attempt to engage occurrences of beings out of originary temporality alone and through a transcendental articulation. But in this failure Heidegger's thought, which ultimately leads beyond its event, lets the alterity and exilic character of thought begin to be heard. In other words, in the ephemeral enactment of Heidegger's thought philosophical discourse begins to figure a re-turning passage; an echo that, like the beat of a ship's bow against the waves in the night, lights the heavens by which it travels and sounds the depths below in its passage. This same passage issues forth a call, a voice, the mark of a passing traveler toward foreign shores: "*Stimmen, nachtdurchwachsen, Stränge / An die du die Glocke hängst* (Voices veined with night, ropes / you hang the bell on)."[33] If something may be said to have been accomplished by this discussion, it is the distinct sense of a need to begin again, to begin to learn to engage thought with a voice that will let both thought's concrete yet ephemeral passages and its alterity be experienced. And, with this open task, one has also encountered the difficulty of understanding, undergoing, and withstanding that need in its essential occurrences.

32. I take Irigaray's *The Forgetting of Air in Martin Heidegger* (Austin: University of Texas Press, 1983) to enact a similar interruption. In both cases thought is recalled to alterity and exilic grounds.

33. Paul Celan, "Sprachgitter" (Speech-Grille), 1959, in *Selected Poems and Prose of Paul Celan,* tr. John Felstiner (New York: W. W. Norton, 2001), 88.

PART THREE

Fugue

Reachable, near and not lost, there remained in the midst of the losses this one thing: language. It, the language, remained, not lost, yes in spite of everything. But it had to pass through its own answerlessness, pass through frightful muting, pass through the thousand darknesses of deathbringing speech. It passed through and gave back no words for that which happened; yet it passed through this happening. Passed through and could come to light again, "enriched" by all this.

—Paul Celan, "Speech on the Occasion of Receiving the Literary Prize of the Free Hanseatic City of Bremen"

7

Concrete Passages

Alterity and Exilic Thought in Heidegger's Later Work

Introduction

The preceding series of encounters with Heidegger's thought in *Being and Time* indicates that this work may be read as a double passage: Heidegger's book may be read first as the beginning of a transformative reappropriation of metaphysical and transcendental traditions; and second as a work that in its passage overcomes its very event, as it leads Heidegger toward another path of thought. In this double passage, Heidegger's thought enacts its alterity and points to its exilic character. These aspects of *Being and Time* are engaged and traced through the way in which the figure of spatiality is operative in Heidegger's discourse on the originary temporality of events of beings and of thought. The following questions arise out of these encounters, and introduce this chapter as a point of closure to the engagements with the alterity and exilic character of Heidegger's thought in *Being and Time*: Does Heidegger's later thought engage its alterity and exilic character? And if so, how does this engagement occur?

The self-overcoming character of Heidegger's thought in *Being and Time* leads to his retraction of the third part of that work, to his abandonment of his attempt to ground dasein's spatiality in temporality, which happens in *On Time and Being*, and ultimately, to his famous "turn" (*die Kehre*). These moments also point to Heidegger's turn toward understanding philosophical discourse as a series of paths of engagement with the occurrences or events of beings. This is explicit in some of the titles of his later works: for example, *Holzwege* (Forest Path), and *Wegmarken* (Pathmarks).[1] Among the various paths of Heidegger's later thought spatiality reoccurs increasingly and at crucial moments in the engagement of events of beings and thought. Alongside this issue appears the alterity and exilic character of thought. Three crucial works that bare such issues in Heidegger's later thought are *Contributions to Philosophy* (1936), "The Origin of the Work of Art" (1930–35), and *Art and Space* (1969).

Contributions to Philosophy: Enactments of Alterity and Exilic Thought

In his second major work, *Contributions to Philosophy*,[2] Heidegger's thought engages the disclosedness of events of beings and thought in their concrete finitude by thinking these events out of the time-space (*Zeit-Raum*) that belongs to being's essential sway (*Wesung des Seins*), or in other words, that belongs to the disclosure or presencing of beings and thought.[3] This means that Heidegger seeks the disclosure of beings in

1. *Holzwege, GA* 5; *Wegmarken, GA* 9; and *Aus der Erfahrung des Denkens. GA* 13, respectively.

2. *Beiträge zur Philosophie (Vom Ereignis), GA* 65; *Contributions to Philosophy*, tr. Parvis Emad and Kenneth Maly (Indianapolis: Indiana University Press, 1999). Unless noted otherwise, quotes are from this translation, cited hereafter as *Contributions.*

My interpretation of Heidegger's *Contributions* follows the many helpful indications and the scholarship of the contributors to *A Companion to Heidegger's Contributions*, ed. Charles E. Scott, Susan Schoenbohm, Daniela Vallega-Neu, and Alejandro Vallega (Indianapolis: Indiana University Press, 2001); as well as Daniela Vallega-Neu, *An Introduction to Heidegger's Contributions* (Indianapolis: Indiana University Press, 2003).

3. "Der Zeit-Raum als der Ab-grund," in *Beiträge zur Philosophie (Vom Ereignis), GA* 65, 371–88; *Contributions*, 259–71. See also Friedrich-Wilhelm von Herrmann, "Wahrheit-Zeit-Raum," in *Die Frage Nach der Wahrheit*, ed. Ewald Richter (Frankfurt am Main: V. Klostermann, 1997). Specifically, see Part III, "*Wahrheit des Seyns und Zeit-Raum in der seinsgeschichtlichen Fragebahn*," 249–56.

terms of temporality *and* spatiality. How does this occur? *Contributions* is meant to be a passage, from what Heidegger calls the epoch of going over from the first beginning of Western philosophy (metaphysics, transcendental philosophies, pragmatism) to the other beginning of philosophical thought.[4] The task of his book is therefore to prepare or open a way for the other beginning, i.e., for a beginning that in part occurs in the engagement of the occurrences of beings and thought in their concrete finitude and temporality. This preparatory thought occurs as an event that leads to the engagement of the essential sway of the occurrences of beings and thought in their concrete disclosedness (*Wesung des Seins*). At the same time, according to Heidegger, in its event this preparatory thinking is meant as an enactment (*Vollzug*) that figures the possibility of the engagement of being's essential sway.[5] Heidegger specifies this when he writes that *Contributions* occurs as an enacting thinking (*erdenken*).[6] The time-space play and the enacting thinking of *Contributions* hold open the possibility of the essential sway of being—which is to say that what the enacting thinking prepares is given in the preparing thought. The crucial point is that in enacting a preparatory thinking, the thinking of *Contributions* does not look ahead to something other, to a horizon. Rather, it is the occurrence of this thinking as such, in its enactment, that *is* the time-space play in which being's essential sway is disclosed in its possibility. In Heidegger's words, "[t]his saying does not stand over against what is said. Rather, the saying itself *is* the 'to be said,' as the essential swaying of beyng."[7]

The enacting or performative character of the thinking of *Contributions* can be seen more clearly in contrast with the transcendental elements of Heidegger's articulation of the occurrences or events of beings and thought in *Being and Time.* What marks the difference between

---

By "concrete finitude" I mean to indicate the temporality of the occurrences or events of beings and thought in their futurity, withdrawing, loss, alterity, and exilic aspects, as already discussed in the previous chapters. Also, the lengthy phrases that accompany the German terms are meant as engagements of Heidegger's thought (as indicated in German), and not as a literal translation from the German.

4. *GA* 65, 3 (*Contributions,* 3).

5. *GA* 65, 5: "[D]ie gründende Eröffnung des Zeit-Spiel-Raums der Wahrheit des Seyns" (*Contributions,* 4).

6. *GA* 65, 5 (*Contributions,* 4, my translation: "*Contributions* question along a pathway. . . . This pathway brings the crossing into the openness of history and establishes the crossing as perhaps a very long sojourn, in the enactment of which the other beginning of thinking always remains only an intimation, though already decisive").

7. *GA* 65, 13: "[D]en Zeit-Raum der letzten Entschseidung . . . vorzubereiten" (*Contributions,* 10).

Heidegger's first work and *Contributions* is precisely the absence of a transcendental horizon and of a dasein that must transcend itself to reach such a temporal horizon. In the latter work, Heidegger thinks not toward temporality but out of the temporal occurrence of the sway of being as such.[8] Unlike in *Being and Time,* the issue is not a certain overcoming or a move toward a temporal horizon, but instead thinking sets out from the occurrence of being's essential sway in its possibility, i.e., out of both the very events of beings and thought in their concrete finitude and temporality. Heidegger calls this thinking a thinking "from enowning" (*vom Ereignis*). This engagement of events of beings and thought is marked in *Contributions* by Heidegger's use of the archaic spelling of *Sein* as "*Seyn,*" which I will mirror by using the term "beyng." The significant point for us is that in *Contributions* the philosophical discourse is the enactment of events of beings and thought in their possibilities. In other words, the discourse of *Contributions* is the unique (*einzige*) opening occurrence and possibility of the time-space play of the essential sway of beyng.[9] This last statement becomes clearer when one turns to Heidegger's understanding of the possibility of the engagement/enactment of events of beings and thought, to the possibility of beyng's essential sway in *Contributions.*

According to Heidegger, the thought of *Contributions* is a preparation that in its occurrence enacts and therefore holds open the possibility of the arising of the events of beings and thought (beyng) in their decision (*Entscheidung*).[10] This last term indicates the concrete and difficult occurrence of the thought of *Contributions* in its finitude or epochal situatedness. In §44, in which Heidegger discusses decision, it becomes clear that *Entscheidung* indicates neither a decision to be made, a choice made by a subject, consciousness or will, nor a decision that stands ahead of the thinking of *Contributions.* In this section Heidegger repeatedly designates decision as the "or": in other words, the in-between that places the thinking of *Contributions.* I say "places" because the possibility of the arising of events of beings and thought (beyng's essential sway) in decision is thought out of the unique taking place of thinking in our epoch,

8. *Contributions,* 4, italics in original.

9. See von Herrmann, "Wahrheit-Zeit-Raum," in *Die Frage Nach der Wahrheit.*

10. *GA* 65, 13: "Das Seyn aber ist nicht ein "Früheres"—für sich bestehend—, sondern das Ereignis ist die zeiträumliche Gleichzeitigkeit für das Seyn und das Seiende" (*Contributions,* 10: "But beyng is not something earlier—subsisting for and in itself—, rather enowning is the temporal-spatial simultaneity for beyng and beings").

i.e., in the event of thought in between the first (metaphysics) and other beginning for thought, and in its enactment of this passage out of its very finitude. This thinking in between or in hesitation is what Heidegger indicates from the very preview, in which he speaks of the thinking of *Contributions* as a thinking on the way (*Gedanken-gang*),[11] a thinking-saying-in-between (*Zwiesprache*),[12] a thinking in the epoch of going over (*Übergang*).[13] In *Contributions* the time-space play of beyng's disclosure occurs, is enacted, in the rigorous exercise of remaining with (withstanding, undergoing, understanding) the occurrences of beings and thought in our epoch, which according to Heidegger is the epoch of the abandonment of beings by beyng. Thinking in our epoch, understanding, withstanding the passing sway of beyng, does not mean looking ahead or being led to what is already there. Rather, it means remaining and undergoing the ephemeral passage of thought as it enacts and opens the essential occurrence of beyng to its possibility.[14] By remaining in decision, i.e., with the concrete finitude of thought in its epoch or taking place, thought enacts the arising and passage of events of beings in the concrete finitude of their possibilities or presencing. This abiding in the finitude of events of beings and thought is an understanding that requires the attunement of thinking to the passing of beyng. According to Heidegger, this attunement attunes thinking through dismay (*Erschrecken*), reservedness (*Verhaltenheit*), and *Scheu*.[15] The last word is generally translated as "awe," but it indicates also silence, delicacy, and vulnerability.

Figured in these modes of attunement is Heidegger's direct engagement of the alterity and exilic character of thought and the occurrences or events of beings. It is not through a re-presentational discourse about beings that occurrences of beings and thought are engaged. Thought's enactment of its finitude opens the possibilities for the engagement of all possibilities of beings and thought in their arising and passage. Terms of attunement such as dismay, reservedness, and the silence and vulnerability

11. "Die 'Entscheidung,'" in *GA* 65, 90–103 (*Contributions,* 62–72). Cf. "Für die Wenigen," in *GA* 65, 11 (*Contributions,* 9).

12. *GA* 65, 3 (*Contributions,* 3).

13. *GA* 65, 6 (*Contributions,* 5).

14. *GA* 65, 3 (*Contributions,* 3).

15. *GA* 65, 13: "Das Sagen von der Wahrheit; denn sie ist das Zwischen für die Wesung des Seyns und die Seiendheit des Seienden. Dieses Zwischen gründet die Seiendheit des Seienden in das Seyn" (*Contributions,* 10: "For Truth is the between [*das Zwischen*] for the essential swaying of beyng and the beingness of beings. This between grounds the beingness of beings in beyng").

of *Scheu* figure neither a representational approach nor an objective and ideal interpretation of the phenomena and thought. Although these terms are the fundamental modes of understanding in *Contributions,* they do not point to presence or representation.[16] Reservedness, dismay, awe: these terms indicate the finite, ephemeral, and concrete character of thought. At the same time, what will attune thinking to beyng's essential sway is not "awe" in the face of some other being, nor in the absence of "someone" or "something" that must remain beyond, a *mysteria.*[17] These attunements open possibilities by remaining with the arising and coming to pass of the event of thought. In other words, these modes engage the very event of thought in its passing, and through such attunement perform the opening for the essential sway of beyng. But how do the alterity and the exilic character of thought figure in Heidegger's *Contributions*?

According to Heidegger, in *Contributions,* events of beings and thought arise in the enactment of a certain time-space play. This time-space play arises as an event that is neither the "ground" for the occurrences of beings nor "grounded" in temporality. Rather, it engages temporality in its concrete passage. This leads Heidegger to state that ground only occurs as an "ungrounding," or abyssal event (*als Abgrund*): "Abyss [*Abgrund*] is the originary swaying of ground."[18] In other words, the occurrences or events of beings and thought arise in the enactment of their events of finitude, i.e., in their coming to pass in presencing or, alternately, in their arising to presence in light of their loss. When Heidegger speaks of "ground" (*Grund*) as ungrounding (*als Abgrund*) in *Contributions,* he means the finitude of events of beings and thought that figures the very passing of their events or presencing. However, unlike in *Being and Time,* here the structure of temporality is engaged not in terms of the temporal horizon that grounds the presencing of events of beings and thought, but rather by thinking out of the withdrawing aspect of temporality—which is operative in all presencing of beings and thought—and by performing that withdrawing in the passage of the thought in *Contributions.* Herein, Heidegger engages the presencing of events of beings *als Abgrund* in a double motion. The ungrounding of beyng occurs as *Entrückung* and *Berückung.*[19] *Entrück-*

16. *GA* 65, 14 (*Contributions,* 11).

17. "The grounding-attunement of thinking in the other beginning resonates in attunings that can only be named in a distant way." *Contributions,* 14.

18. This is also the case for religious thought. See "The Last God," in *Contributions,* 405–17.

19. *Contributions,* 379. See also "Der Zeit-Raum als der Ab-grund," in *GA* 65, 371–88 (*Contributions,* 259–71).

ung refers to the withdrawing aspect of beyng, whereas *Berückung* refers to the rising or coming out that occurs in that withdrawal. As already indicated, this engagement sets out from temporality now rethought in the unfolding of the time-space of the truth of beyng and not toward temporality; this occurs by thinking in the withdrawal and absencing of events of beings and thought in their concrete finitude. It is out of this withdrawal or absencing that presencing is thought. Or, put another way, the withdrawing figures the ground of all presencing as ungrounding, since beyng's essential sway occurs out of its own withdrawal. In short, absencing, withdrawal, and loss figure the ground for all presencing, and therefore appear as the necessary ungrounding of beings and thought.

Alterity and exilic thought do not, then, appear as the horizon of events of beings and thought, as is the case in *Being and Time*'s engagement of temporality and finitude, but are understood and engaged as arising out of withdrawal and loss, and as enacting such passages in their coming to pass. In short, *Contributions* thinks out of alterity and in the enactment of a transformative openness in ephemeral passage. This is clear when one looks at what has already been discussed concerning the character of the thought of *Contributions* as enacting thinking in light of this ungrounding sense of temporality. As already indicated, the thinking of *Contributions* arises in light of the withdrawal of beyng, and thus it arises and also comes to pass out of a certain alterity. But this thought also figures an exilic event, in that it occurs as a thinking in between, and in that it enacts in its ephemeral passage the possibility for the engagement of beyng by remaining on exilic grounds, by remaining with decision in its epochal taking place, in its finitude. In other words, it figures an event that cannot be grounded on ever-present origins, and that looks ahead to open possibilities of beings and thought only intimated in the phrase "other beginning," but figured by thought's finitude.

"The Origin of the Work of Art": The Sheltering of Alterity and Exilic Thought

In his lecture "The Origin of the Work of Art" (1935), Heidegger discusses the sheltering of the truth of being in beings.[20] The "truth of

20. *GA* 65, 385 (*Contributions*, 268). Cf. *Besinnung, GA* 66, 101–2.

being" here figures the absencing/presencing that is operative in events of beings and thought. (This issue in Heidegger's thought was indicated in Chapter 2 as an elemental characteristic of his understanding of language and the possibility of engagement of events of beings and thought, and is figured by the Greek word *aletheia.*) When one looks at the essential motion of Heidegger's thought in "The Origin of the Work of Art," one finds an articulation of the essential sway of beyng in terms of the concrete finitude of events of beings and thought.[21] Heidegger's lecture engages the "sheltering" of this essential sway in its presencing and absencing or withdrawal (the truth of beyng) in the concrete finitude of events of beings and thought. The issue in this work is specifically the thinking of the sheltering of the truth of being, i.e., the way finite, concrete occurrences of beings and thought figure the absencing or withdrawal operative in all events of presencing. Heidegger's discussion is particularly concerned with the work of art. The following words from the introduction are useful in order to introduce Heidegger's project: "What something is as it is we call its essence. The origin of something is the source [*Herkunft*] of its essence. The question concerning the origin of the work of art asks about its essential source."[22] The question of "origin" does not ask after things at hand, either as material objective presence, or as ever-present, unchanging metaphysical or transcendental essence. This question is presented from the start by Heidegger as the question of the source (*Herkunft*) of the thing and its essence.[23] This term indicates what is at issue in the passing of events of beings and thought in their finitude (absencing/presencing), i.e., the source event of the occurrences of beings and thought. As we have seen, this source is the coming

21. "The Origin of the Work of Art," in *Basic Writings,* ed. David Farrell Krell (San Francisco: Harper, 1993) (hereafter referred to as "Origin"); *Der Ursprung des Kunstwerks, GA* 5, with *Nachwort* and *Zusats* from 1960. Cf. "Vom Ursprung des Kunstwerks. Erste Ausarbeitung (1931/32)," *Heidegger Studies* 5 (1989): 5–22; "Zur Überwindung der Aesthetik. Zu 'Ursprung des Kunstwerks' (1934)," *Heidegger Studies* 6 (1990): 5–7. For a discussion of the series of compositions that lead to the final lecture see Francoise Dastur's thorough study, see "Heidegger's Freiburg Version of the Origin of the Work of Art," in *Heidegger Toward the Turn,* ed. James Risser (Albany: SUNY Press, 1999).

22. The writing of this lecture directly overlaps Heidegger's writing of *Contributions*: the lecture in its final form is dated 1935, *Contributions* is dated 1936. However, the transformation marked by the spelling "beyng" does not appear in "The Origin of the Work of Art." Although I move from beyng to being, I take the difference in spelling to mark an opening for questioning the lecture's relation to *Contributions* and its place within that work.

23. *GA* 5, 1: "Das, was etwas ist, wie es ist nennen wir seines Wesens. Der Ursprung von etwas ist die Herkunft seines Wesens. Die Frage nach dem Ursprung des Kunstwerkes fragt nach seiner Wesensherkunft" ("Origin," 143).

to pass of events of beings and thought engaged out of the absencing or withdrawing essential to their events. But how are the occurrences or events of beings and thought engaged in their concrete finitude and out of the withdrawal of being?

Heidegger writes in "The Origin of the Work of Art":

> To be a work means to set up a world. . . . The world is not the mere collection of the countable or uncountable, familiar or unfamiliar things that are at hand. . . . The world worlds [*die Welt weltet*], and is more fully in being than the tangible and perceptible realm in which we believe ourselves at home. World is never an object that stands before us and can be seen. . . . By the opening up of a world, all things gain their lingering and hastening, their remoteness and nearness, their scope and limits. In a world's worlding is gathered that spaciousness out of which the protective grace of the gods is granted or withheld.[24]

He then goes on:

> A work, by being a work, makes space for that spaciousness. "To make space for" means here especially to liberate the free space of the open region and to establish it in its structure. This installing occurs through the erecting mentioned earlier. The work as work sets up a world. The work holds open the open region of the world.
>
> *Indem ein Werk ist, räumt es jene Geräumigkeit ein. Einräumen bedeutet hier zumal: freigeben das Freie des Offenen und einrichten dieses Freie in seinem Gezüge. Dieses einrichten west aus dem gennanten Er-richten. Das Werk stellt als Werk eine Welt auf. Das Werk hält das Offene der Welt offen.*[25]

According to Heidegger the possibility of the engagement of the occurrences or events of beings and thought in their withdrawing or absencing and presencing occurs through the work. The work in its working sets apart earth and world,[26] and it is in this strife between earth and world

24. In the first paragraph of the introduction, Heidegger writes: "Der Ursprung von etwas ist die Herkunft seines Wesen." See also *GA* 5, 1. "Source" here does not refer to anything other than the sheltering occurring of beings out of the withdrawing of being.

25. *GA* 5, 30 ("Origin," 170).

26. Ibid.; *GA* 5, 31 ("Origin," 170).

that the "world worlds," thus opening the space for beings and thought to shine forth.[27] The brief outline in this passage marks the steps in the unfolding motion of the sheltering of events of beings and thought in their withdrawing and coming to presence: it is in light of the work in its ephemeral passage that anything like world, entities, and thought will come to pass.

The most obvious surprise in this text, and in contrast to *Being and Time,* is that here the absencing/presencing of events of beings is articulated in terms of spatiality, and not only out of temporality. As is generally the case with interpretations of this work, the urgency or overeagerness of the interpreters to adopt ontological concerns (in this case the ontological character of the work, as well as the controversial question of whether this lecture is to be understood purely ontologically or as the beginnings of Heidegger's aesthetics) has overshadowed the distinctly spatial language of Heidegger's articulation.[28] As Heidegger declares, the work "opens" (*öffnet*) a world, a world in its worlding. It is by the opening of this world that things gain their nearness and remoteness (*Nähe und Ferne*). At the same time, in the worlding of this world gathers a certain "spaciousness" (*Geräumigkeit*). This word refers to the gathering (*ge*) and opening (*räumigkeit*) possibility of events of beings and thought in their coming to pass. According to Heidegger the work accomplishes this opening by "making space" (*einräumen*) for the gathering of events of beings and thought in their disclosedness. This making space means "to liberate the free space of the open region and to establish it in its structure" (*freigeben das Freien des Offenen*). In setting up a world in its worlding, the work sets free the possibility for the coming to pass of all events of beings by enacting, sustaining, and withstanding "the openness of the world" (*das Offene der Welt*).[29] The undergoing or passage of

27. John Sallis makes this point most eloquently: "Art is the taking place of truth, a way truth happens." See *Stone* (Indianapolis: Indiana University Press, 1994), 109. This is one of Sallis' central works on the question of the work of art and a clear reflection on the possibility of understanding the work of art after Heidegger. See also Sallis, *Shades—of Painting at the Limit* (Indianapolis: Indiana University Press, 1998).

28. "In setting up a world and setting forth the earth, the work is an instigating of this strife." "Origin," 175. At the same time this strife is the sheltering of the truth of being. Through the opening of world, "the work lets the earth be an earth." "Origin," 172. The sheltering occurs as a twofold passage set in motion by the work, since the work of art sets up a world and in doing so returns to the earth.

29. On the world-earth strife see Michel Haar, "La relation monde-terre chez Heidegger," in *L'oeuvre d'art* (Paris: Hartier, 1994). Cf. Dastur, "Heidegger's Freiburg Version of the Origin of the Work of Art."

events of disclosedness of world, beings, and thought enacts the sheltering of their finitude. The passage, and indeed this section of the lecture, offers a spatial articulation of the motion of the sheltering of events of beings and thought engaged in light of their concrete finitude, or coming to pass.

This does not mean that the sheltering of the truth of being—the sheltering of events of beings and thought in their arising in light of their withdrawal and loss—is "in beings." The events of beings and thought occur as and in the possibility of the worlding of the world, and this opening or disclosedness occurs in the shining forth of all beings in their coming to pass. In his later work, Heidegger also refers to this coming to pass as *phusis.*[30] Beings arise out of their finitude and enact this finitude and arising in their concrete "taking place." Therefore, one can say that in their coming to pass beings shelter the truth of being (the absencing/presencing from which beyng occurs) because they arise out of this truth. Alternately, beings and thought arise, have their source, in their concrete finitude (or temporality), and, if understood in this way, they hold open the possibility of the sheltering of the essential sway of being by rising in light of their finitude. This is what Heidegger indicates when he writes that "[t]ruth happens only by establishing itself in the strife and the free space opened up by truth itself."[31]

This last remark requires a further observation, for the sake of following Heidegger's engagement of the occurrences or events of beings and thought out of their withdrawing and loss or finitude. Heidegger asserts in "The Origin of the Work of Art" that the occurrence of events of beings in their concrete finitude (*Wesen des Wahrheitsgeschehens*) "is historical in multiple ways" (*ist in mannigfaltigen Weisen geschichtlich*).[32] Here, as in *Being and Time* and *Contributions, geschichtlich* refers to the

30. See Otto Pöggeler, *Die Frage nach der Kunst. Von Hegel zu Heidegger* (Freiburg: München, 1984); *Über die moderne Kunst: Heidegger und Klee's Jenaer Rede von 1924* (Jena: Erlangen, 1995); and "Heidegger und die Kunst," in *Martin Heidegger: Kunst. Politik, Technik,* ed. Jamme and Harries (Frieburg: München, 1992): 58–84. See also Friedrich-Wilhelm von Herrman, *Heideggers Philosophie der Kunst: Eine systematische Interpretation der Holzwege-Abhandlung 'Der Ursprung des Kunstwerkes'* (Frankfurt am Main: V. Klostermann, 1980).

31. The translator uses "region" for *das Offene,* but this can be confused with *Gegend,* which literally means "region," and marks a different aspect of the disclosedness of being. See "Die Kunst und der Raum," in *Aus der Erfahrung des Denkens, GA* 13, 203–10.

32. "The Greeks early called this emerging and rising in itself in all things phusis." "Origin," 168. On *phusis,* see "Aletheia (Heraklit Fragment 16)," in *Vorträge und Aufsätze* (Pfüllingen: Neske, 1985), 163–66; "Aletheia (Heraclitus Fragment B16)," in *Early Greek Thinking,* tr. David Farrell Krell and Frank A. Capuzzi (New York: Harper & Row, 1975), 114. Cf. Heidegger,

occurrence of being in its essential sway, and not to the objective measurable fact of *Historie.*[33] More specifically, *geschichtlich* refers to the *Geschehen* or "occurrence" of the essential sway of the truth of being (*Wahrheitsgeschehen*). At the same time, the sheltering motion of the truth of being is the setting of itself into work, in words, works of art,[34] with humans as creators (*Schaffende*) and preservers (*Bewahrende*),[35] and with thinking (*das Fragen des Denkers*).[36] Out of the essential sway of the truth of being in its self-sheltering motion these configurations of beings find their place. At the same time, place, work, the word, thoughts,[37] and the work of art shelter as they come to be in the truth of being. One might say, for example, that the word shelters the truth of being but that the word cannot place itself in this truth, i.e., the word does not determine events of beings or thought. This points to a certain understanding of beings out of the truth of being. Things, bodies, concreteness, identities, and institutions are to be thought in their essential occurrence, not as objects or essences already given, but out of their source, in their occurrence, and out of the truth of being (*Wahrheitsgeschehen*).

This indicates that, for Heidegger, the occurrences or events of beings and thought enact in their coming to pass the very possibility of their events. This engagement of beings and thought in their passing thus figures the engagement of the alterity and exilic character of such events. In the enacting/sheltering of events of beings and thought "no-thing" is sheltered. Sheltering is not a matter of re-presenting entities at hand or ever-present, unchanging essences or principles. The sheltering of beyng by beings figures the sheltering of the finitude of events of beings and thought in their coming to pass. These events of beings and thought shelter finitude, and in doing so recall the alterity of their events in their coming to presence. Furthermore, this means that events of beings and thought in their sheltering enact a passage that cannot be traced to unchanging origins or principles, and also means that these events remain

---

Heraklit, "Der Anfand des Abendländischen Denkens (Heraklit)," GA 55, 127–31; *Grundfragen der Philosophie. Ausgewählte "Probleme" der "Logik," GA* 45.

33. "Origin," 181.

34. *GA* 5, 49 ("Origin," 186).

35. *GA* 65, 32: "Geschichte hier nicht gefaßt als ein Bereich des Seienden unter anderen, sondern einzig im Blick auf die Wesung des Seyns selbst" (*Contributions,* 23). Cf. *BT,* 17 (*SZ,* 21–20).

36. *GA* 5, 49 ("Origin," 187).

37. *GA* 5, 54 ("Origin," 191). Cf. *GA* 65, 17–18: "*Sucher, Wahrer, Wächter*" (*Contributions,* 13: "Seeker, preserver").

open, in light of their own passage, to transformations yet to come—i.e., in their occurring in light of their finitude or their not-yet. According to this particular lecture, then, the disclosedness of events of beings and thought in their coming to pass can only occur in light of the concrete passage of beings and thought (which, as I have already indicated at the beginning of this section, are not interpreted as entities at hand and essences, but understood and withstood in their ephemeral and historical [*geschichtlich*] passages).[38] Still, the difficulty of the engagement of the alterity and exilic character of the occurrences or events of beings and thought in their concreteness remains only a proposition. Heidegger articulates the truth of being in terms of spatiality in another text, and specifically in the concrete sense of spatiality as the "place" (*Ort*) of the withdrawing and presencing of events of beings and thought.

Art and Space: Embodying Thought's Alterity and Exilic Grounds

Heidegger's 1969 work *Art and Space*[39] discusses the sheltering of events of beings and thought in their concrete finitude, a concreteness marked by the fact that Heidegger literally writes this piece on stone.[40] In this work, the sheltering of the truth of being is again engaged through the work of art and in terms of spatiality. This time the discussion focuses on

38. "Origin," 187; *GA* 5, 49.

39. Cf. *GA* 65, 12 (*Contributions,* 10: "It is only through the ones who question that the truth of beyng becomes a distress [*Not*]").

40. At this point of engagement with events of beings and thought in light of their concrete coming to pass, Heidegger's lecture points to a powerful question. Because of the finitude and ephemeral character of events of beings and thought, and in view of their alterity and transformative/exilic characters, all ideologies, political agendas, the "facts" and "realities" of life, and their necessary call to action, cannot be taken as simply given concepts, themes, problems, and programs that must be followed as directives for thought. In light of the alterity and exilic character of thought and events of beings, these configurations of life raise hermeneutic rather than a "political" questions: How will thought engage such configurations of the occurrences or events of beings and thought (ideologies, themes, etc.) in a way that remains with their alterity and transformative possibilities? How will thought and action shelter the absencing and withdrawing operative in all presencing, such that the sheltering of the occurrences or events of beings and thought in their possibilities may be enacted? Understood from the perspective of the task of the sheltering of being in its finitude and the possibilities intimated by such engagement, the "realities" of life turn from "fact" to passages that remain well beyond fact, ideology, measurement, calculation, implementation, and ultimately presence, objective or ideal.

sculpture (*Plastik*). Briefly, according to Heidegger, the spatiality of the work of art, in bringing-into-the-work the truth of being, occurs as a sheltering, which he calls a "clearing away" (*Räumen*). This clearing away "brings forth the free, the openness for man's settling and dwelling."[41] It is in this clearing away that the work enacts the sheltering of events of beings and thought in their finitude. This figure of the spatiality of the disclosedness of events of beings is analogous to the spaciousness of the worlding of the world in "The Origin of the Work of Art."[42] Heidegger then goes on to write that "[c]learing is releasing the places at which a god appears."[43] The clearing away, *Räumen,* refers to the very event of disclosure of the truth of being. This disclosure figures a performative event, because the occurrences or events of beings come to pass as a making room (*Einräumen*).[44] Just as in "The Origin of The Work of Art," the discussion refers here to the enactment of the possibility of the occurrences or events of beings and thought in their concrete coming to pass. This enactment is a twofold motion out of which, on the one hand, "openness holds sway" (*es läßt Offenes walten*) for the shining forth of things, and on the other hand, beings find their relation.[45] In this twofold openness "places happen."[46] Heidegger concludes in a single-sentence paragraph: "Clearing is releasing of places" (*Räumen ist Freigabe von Orten.*)[47]

The sheltering of the occurrences or events of beings and thought occurs as the granting of places (*Ort*). Heidegger maintains that "[p]lace always opens the region in which gather things in their belonging together. Gathering comes to play in the place in the sense of the releasing sheltering of things in their region."[48] Place shelters and gathers beings in the truth of being. Here, place is conceived neither in terms of objective presence at hand nor in terms of the beingness (*Seiendheit*) of beings, nor is

41. *Die Kunst und der Raum* (St. Gallen: Erker-Presse, 1969); also published in *Aus der Erfahrung des Denkens*, *GA* 13. See also *Art and Space,* tr. Charles H. Seibert, in *Man and World* 9 (1973), 3–8.

42. *GA* 13, 250.

43. *GA* 13, 206: "Das Räumen erbringt das Freie, das Offene für ein Siedeln und Wohnen des Menschen" (*Art and Space,* 5).

44. See the discussion in the previous section.

45. *GA* 13. Cf. *GA* 5, 31 ("Origin," 170: "In a world's worlding is gathered that spaciousness [*Geräumigkeit*] out of which the protective grace of the gods is granted or withheld").

46. *GA* 13, 207: "Wie geschiet das Räumen? Ist es nicht Einräumen . . . ?" (*Art and Space,* 6).

47. Ibid.

48. *GA* 65, 207: "Im zwiefältigen Einräumen geschieht die Gewährnis von Ort" (*Art and Space,* 6).

it an ideal space outside of the taking place of the work. Heidegger is clear about this: "Place is not located in a pre-given space, after the manner of physical-technological space. The latter [space] unfolds itself through the reigning of places of a region."[49] Instead, places arise as the clearing away and making room, which is the sheltering enacted by events of beings in their concrete finitude or coming to pass. In this sense, one may say that in their occurrences places hold open the possibility of the truth of being in its essential sway. In contrast to the thinking of *Being and Time,* here spatiality is not engaged in terms of entities, but rather has the performative character of letting beings be. Spatiality appears as a figure similar to Timaeus' *chora,* a kind beyond kind, a figure of the presencing of events of beings that remains outside determination in terms of objective and ideal presence. However, the coming to pass of events of beings and thought finds its possibility and sheltering in the concrete taking place of beings. Heidegger writes that "things themselves are places."[50] Thus, the sheltering of the truth of being in places will require some understanding of the occurrence of beings in their configurations as things, and in a way other than the objective and ideal dualism of traditional metaphysics and transcendental thought. At this point the question of sheltering and gathering must turn to that of embodying.[51]

In *Art and Space* Heidegger speaks of sculpture as work: "Sculpture: an embodying [*verkörpendes*] bringing-into-work of places, and with them a disclosing of regions of possible dwelling for man, regions for the possible tarrying of things surrounding and concerning man. Sculpture: the embodiment of the truth of being in its work of instituting places."[52] Works of art, words, thoughts (in this case sculpture), all embody the truth of being by their concrete taking place. Sheltering in its taking place does not concern "place" as an abstract concept: sheltering occurs in the taking place of beings and thought. This indicates at least two clear aspects of Heidegger's thought. First, his engagement of the sheltering of the truth of being does not leave behind or forget the concreteness of beings in their finitude or coming to pass. Second, this concreteness must be thought in terms of the task of sheltering the truth of being, i.e., by engaging events of beings and thought out of their absencing/presencing, and understanding their concreteness in light of this—their finitude and

49. *GA* 13, 207 (*Art and Space,* 5).
50. Ibid.
51. *GA* 13, 208 (*Art and Space,* 7).
52. Ibid.

ephemeral passages, their alterity and exilic character. To put it another way, what is to be thought is the embodying/sheltering of beings in their coming to pass.[53]

According to *Art and Space,* embodying is a matter of enacting the concrete finitude of events of beings engaged out of their withdrawing or absencing, in light of which they come to presence. This means, first of all, that the occurrences or events of beings must be engaged out of their withdrawing or absencing, and second that they must be understood not merely as entities at hand, but as enactments of such withdrawal and loss. Understood in light of the loss operative in their presencing, beings and thought cannot be interpreted in terms of objective and ideal presence, nor through the traditional representational thought sustained by such interpretations. As already indicated in the first part of this book, the coming to pass of beings and thought cannot be engaged by way of such thought. This is what Heidegger means when he writes of the way sculpture can be engaged: "What is named by the word 'volume,' the meaning of which is only as old as modern technological natural sciences, would have to lose its name."[54] Works (again in this case sculpture) will have to twist free from interpretations in terms of objective and ideal volume, and one can add the measurable presence and the representational and calculative thought that traditionally name them. At this point the figure of the alterity of the occurrences or events of beings and thought in their passage in loss appears; and with it, unlike in Timaeus' likely story, an attempt to remain with this alterity and in the engagement of the concreteness of events of beings and thought. At the same time, the issue of the name also refers to the exilic character of these occurrences.

Heidegger states in the same section of this work that "t]he place-seeking and place-forming characteristic of sculptured embodying would first remain nameless." This indicates a direct engagement of the exilic character of beings and thoughts. The concrete and embodying enacting of all such events either understood, withstood in their withdrawing, or coming to pass, occurs outside presence and, more specifically, outside the name. There will not be a name, a category, a representation of the occurrences or events of beings. There will not be a naming of beings either as entities present at hand or in terms of ever-present, unchanging origins, principles, or essences that will give beings and thought their

53. I have translated *Verkörperung* as "embodiment," and *verkörpern* as "embodying." Both words refer to the sheltering of the truth of being.

54. *GA* 13, 209–10 (*Art and Space,* 8).

identity and sense. The sheltering of beings and of thought in their coming to pass will be enacted in remaining with and withstanding their ephemeral passages as such. But such engagement of beings will not occur in terms of "thinghood," entities at hand. Rather, it will occur in the engagement of beings in light of the impossibility of tracing their events to ever-present, unchanging origins, in the loss of names. In this way will thought enact the sheltering of events of beings and thought in their concreteness. A last return to Heidegger's text on sculpture indicates the full exilic aspect of his discussion. In light of the break from objective and ideal presence as the way to interpret the occurrences or events of beings, Heidegger asks "[a]nd what would become of the emptiness of space?" He then replies "[e]mptiness [*Leere*] is not nothing. . . . In sculptural embodying, emptiness plays in the manner of a seeking-projecting instituting of places."[55] Beings in their coming to pass are not "nothing." They are temporal/spatial occurrences, passages that figure their temporality not only by arising in an absencing/presencing movement, but in letting their events broach their possibilities by enacting their passage as a passing. The emptiness of space offers possibilities neither objectively nor ideally present, possibilities intimated in the futurity of events of beings and thought. Embodying finds its most powerful aspect as emptiness, as the enactment of a passage that in its presencing already figures a transformative event that will lead beyond its configuration. The occurrences of beings and though enact the sheltering of their passages in events beyond ever-present origins, and as figures of possibilities yet to come. In other words, beings and thought come to pass in alterity and in the enactment of their exilic grounds.

Clearly, this conclusion has moved beyond Heidegger's intention and theme in *Art and Space.* However, by tracing his thought in this piece as well as in "The Origin of the Work of Art" and *Contributions,* one begins to understand the difficult task of engaging the alterity and exilic grounds of thought and beings. For Heidegger these insights are heard in their most distinct form with respect to the ways in which poetry engages the

55. This question of embodying is already present in Heidegger's thought during the period of *Being and Time.* See *Metaphysische Anfangsgründe der Logic im Ausgang von Leibniz, GA* 26, 173–74. Here Heidegger speaks of *Leib* as a fundamental element in the possibility of dasein. The question of embodying is also to a certain extent present in "The Origin of the Work of Art." See also Sallis, "Temple of Earth, "in *Stone.* This issue of thinking in the sheltering embodying of beyng is also taken up and used effectively as a supplement to Heidegger's thought in Luce Irigaray, *The Forgetting of Air in Martin Heidegger,* tr. Mary Beth Mader (Austin: University of Texas Press, 1983).

occurrences of beings in their finitude. Because of the powerful role of poetry (i.e., words) in his thought, it is fitting that this chapter come to a close with a brief discussion of this aspect of Heidegger's thought.

Of Silent Places and Possibilities

In "The Essence of Language" (*Das Wesen der Sprache*), which discusses the poetry of Stefan George and Friedrich Hölderlin, Heidegger writes that "[t]ime's removing and bringing to us, and space's throwing open, admitting and releasing—they all belong together in the same, the play of stillness. . . . The same, which holds space and time gathered up in their nature, might be called the free scope, that is, the time-space that gives free scope to all things."[56] Beings arise as a play of stillness; they arise out of a silence, which, like emptiness, shelters intimations of possibilities felt in that silence. The question is, how will one begin to engage this stillness, and, with it, one's thought in its event? What could it mean to embody the disclosure of events of beings and thought in their play of stillness? How will thought enact the arising of events of beings in light of their loss, in light of their echoing out of stillness and on to stillness? How will words engage their loss and exilic passages?

Heidegger closes his essay with a reflection on a line from George's poetry:

> In the neighborhood of Stefan George's poem we hear it said:
> *Where words break off no thing may be.*
> Now thinking within the neighborhood of the poetic word, we may say, as a supposition:
> *An "is" arises where the word breaks up.*
> To break up here means that the sounding word returns into soundlessness, back whence it was granted: into the ringing of stillness which, as saying, moves the regions of the world's fourfold into their nearness.
> *This breaking up of the word is the true step back on the way of thinking.*[57]

56. *GA* 13, 209 (*Art and Space,* 7).
57. Ibid.

The last line recalls the need to take a step back to the issue of the truth of being. If the word is a site of the embodiment of this truth, in this passage Heidegger indicates that embodying belongs to the truth of being in its withdrawing from or ungrounding of truth at least in two ways. Not only does embodying call for thinking out of the finitude, futurity, loss, and transformative passage figured by the phrase "truth of being," but also to shelter the truth of being will mean to recall its absencing, its performative passage, to its stillness. To put it in terms of "The Origin of the Work of Art," the embodying occurrence of the truth of being will come to pass by enacting a recalling of its event out of the openness of the world to the earth. Only in the embodying that returns to stillness will the sheltering of the truth of being be enacted. This does not mean that one should stop speaking, abandon words, and thereby, unavoidably, thought. On the contrary, it will be through the embodying enacted by ephemeral words that the sheltering of beings and thought will occur. In *Being and Time* Heidegger already indicates this much, when he writes that silence is only possible in the fullness of the saying of the truth of being. However, as this study has often noted, the sheltering of events of beings and thought will not be engaged in terms of objective and ideal presence, or in the representational and calculative sense of language (which, according to Heidegger, sustains the metaphysical and transcendental traditions). As Heidegger indicates in his reflections on George, the echoing of the word in its returning to a certain break, to a certain absence, and in such return to silence and stillness, recalls the occurrences or events of beings and thought to their concrete finitude and ephemeral passages. In the recalling of beings to stillness the word may engage the disclosure of beings and thought in their finitude. So, Heidegger invites one to make a supposition: "*An 'is' arises where the word breaks up.*" He then indicates the matter to be thought if words and language will be engaged and understood in their sheltering possibilities: the recalling to stillness, the break up of words, echoes and marks a step back to the source of alterity and exilic grounds of thought.[58]

Another way to understand this concrete embodying of finitude is by drawing a contrast between the representational language of objective and ideal interpretations of beings and embodying words. The objective presence of beings as entities at hand is not interrupted by representational

58. "Das Wesen der Sprache," in *Unterwegs zur Sprache* (Pfüllingen: Neske, 1986); "The Nature of Language," in *On the Way to Language*, tr. Peter D. Hertz (San Francisco: Harper, 1971), 106.

thought. Indeed, stories about beings establish the continuity of such objective presence, and eventually give interpretations of all events of beings and even their passage as images of an ever-present, unchanging being (metaphysics). Such functioning of the word eventually makes language seem a productive process of sustenance for the beingness of beings (metaphysics): we can only know what we make. But, according to Heidegger, the words of certain poets echo in their ephemeral passages the finitude of the occurrences or events of beings and thought. Such words echo their finitude by breaking presence, by interrupting presence, and by recalling the phenomena to stillness, to silence. This recalling of silence enacts the sheltering of the finitude of beings in their coming to pass, or absencing/presencing events. Certain poetry echoes this finitude with the ephemeral passage of its words, which shelter the withdrawing operative in all presencing events. The passing of such ephemeral voices re-turns presence to its mortality. It is as if *phusis* and words lost their ability to recall; it is as if they were frozen in the face of their presencing—dazzled by the objective presence of entities at hand, except for the mortal words of poets, which return beings to their originary finitude, to their alterity and exilic events.

For Heidegger the embodying of the truth of beings consists of a number of moments or aspects.[59] As Friedrich-Wilhelm von Herrmann indicates in one of his works on Heidegger and the poets, two of these crucial moments are figured by the words of George and Hölderlin, the two poets of "The Essence of Language."[60] The break in George's words marks the silence or emptiness of our epoch, the epoch of the abandoning of beings by being. At the same time, in taking up the withdrawal of the sense of being, this break already echoes being's essential sway. Like George's poetry, Hölderlin's words occur in this absence. However, the difference is that in the case of the latter, words arise out of the abandoning of beings by being, i.e., as an embodying motion of the concrete finitude and withdrawal of the presencing of beings and thought. According to Heidegger, Hölderlin's works engage the occurrences or events of beings by returning the saying to its originary absencing. This insight points to the engagement of the alterity and exilic character of beings and thought by understanding the poet's word as the enactment of the withdrawing or loss operative in the presencing of beings and thought. While

59. *On the Way to Language,* 108.

60. Clearly, "source" does not indicate a causal relation or teleology if thought and beings find their source in silence, absence, alterity, and exilic transformative passages.

George's words intimate the finitude of beings and thought, Hölderlin begins to engage it by the recalling of his words to their stillness. Furthermore, although Heidegger finds in Hölderlin a more fundamental engagement of beings and thought in their concrete finitude, the poetic word only comes to pass as an indication of this. The poetic word indicates something of thought, but it is not identical with thought. As von Herrmann states, the central issue is not "language" as opposed to things or world, but language as the embodying of events of beings and thought in their concrete finitude, or the "announcing" of the essence of language. For Heidegger the difference between George and Hölderlin is that the latter's words return their event to stillness; they recall the finitude of their event in their passage. When, in the fifth stanza of *Germanien,* Hölderlin speaks of "the flower of the mouth," his words enact the sheltering of events of beings and thought by recalling the word to its double play as utterance and physicality, a doubling that echoes the inseparability of earth and world in the coming to pass of events of beings and thought.[61] Hölderlin's poetry recalls the saying to stillness, to the earth. George, whose saying breaks, and in such interruption intimates what remains to be engaged, does not reach Hölderlin's engagement with language. This discussion of the embodying/saying of poetry is directly connected to "The Origin of the Work of Art," where, as indicated above, Heidegger writes that the disclosedness of events of beings is sheltered in that it occurs as a returning of earth to its essential occurrence. At the same time, this recalling will always enact the withdrawal, loss, alterity, and exilic character of the occurrences or events of beings and thought, since it will always re-turn presence to the earth, word to stillness. Hence the saying will enact the loss of origins and transformative possibilities figured by finitude. In this sense, both poets already begin to engage the alterity and exilic character of beings and thought. The engagement of the issue of alterity and the exilic character of events of beings and thought is explored in a direct manner by Heidegger in his writing on poetry in general, as well as in his various discussions of Hölderlin's work.[62]

61. In *Contributions,* for example, the sections on *der Anklang* (echo), *das Zuspiel* (passing), *der Sprung* (leap), and *die Gründung* (grounding) mark unique moments in the essential embodying of the sway of beyng.

62. I base this part of my discussion on von Herrmann, "Die Blume des Mundes: Zum Verhältnis von Heidegger zu Hölderlin," in *Wege ins Ereignis* (Frankfurt am Main: V. Klostermann, 1994), 246–63.

However powerful Heidegger's engagement with poetry, it is certainly not the case for him that poetry will take the place of thought. Again in Heidegger's words, "this break up of the word is the true step back on the way of thinking." This indicates first of all that for Heidegger poetry is not the same as thought, a recurrent misunderstanding often present in interpretations of his thought in relation to poetry. Poetry opens a way for thought in its enacting a break in the word. Furthermore, this obvious observation only leads to the issue behind the differentiation between poetry and thought. As a consequence of the interruption of the traditional interpretation of language and its representational task, a question must be asked concerning what poetry and thought might mean now. This is a central aspect of Heidegger's long engagement with the issue of poetry and thought, an issue that ultimately figures the need to understand the essence of language, and that with such difficulty moves poetry and thought even further toward issues of presencing and absencing, withdrawal, loss, and alterity. But here once again appears a series of questions that, although arisen out of Heidegger's opening of possibilities for thought, mark the limits of his work and thought: What is the limit of poetry as language, when language figures a re-turning of presence toward the stillness of earth? Echoed in at least one of this question's implications is another issue that points beyond Heidegger's focus on "the poet": In the return to earth, does poetry not spread beyond the poet's page into the ephemeral and diversifying passages of beings great and small? How will language begin to engage this recalling to silence in a way that echoes the earth? And what will such saying be called? Poetry? Language? Thought?

Conclusion

We have seen that Heidegger's thought enacts in its motion and event a certain opening for thinking the occurrences of beings and thought in their alterity and on exilic grounds. The various moments in Heidegger's later works where spatiality appears as a central figure for the disclosure of the occurrences of beings show that his later thought goes on to engage

"'The flower of the mouth': Hölderlin's hint for Heidegger's thinking of the essence of language." *Martin Heidegger: Critical Assessments,* ed. Christopher Macann (London: Routledge, 1992).

the occurrences of beings and thought in their finitude, alterity, and exilic character, and that it does so in light of their concreteness or ephemeral passages. The previous reflections led the discussion from enactment to the work and the work's embodying enactment of finitude. This is an engagement of the concrete finitude of events of beings that, according to Heidegger, is figured in our epoch in the words of certain poets, who in their passing words interrupt the certainty of presence and representational language and in this way open the occurrences of beings, including their events, to their alterity and exilic character. These reflections open a path toward an issue that not only belongs to Heidegger's thought and to the work of the poets discussed, but that remains still to be engaged by us in our time and in light of the alterity and exilic character of thought. This issue is perhaps best figured by a question Hannah Arendt asks in a letter to Heidegger a few months before his death: "Where are we, truly, when thinking: the *topos* of the philosopher in the *Sophist.*"[63]

63. Von Herrmann, "Die Blume des Mundes," 283.

"[W]hen it is viewed from within the proper reflection as it occurs in the horizon of ek-sistence which understands world, when it is viewed from within the thrown projecting that stands in the clearing of the self-projecting world, then 'body and mouth' are part of 'the flowing and growth of earth.'" Friedrich Hölderlin, *Poems and Fragments,* tr. Michel Hamburger (N.Y.: Cambridge University Press, 1980), 405.

Bibliography

Primary Works

Works by Heidegger

From the *Gesamtausgabe* (Frankfurt am Main: V. Klostermann, 1975–).

GA 3 *Kant und das Problem der Metaphysik* (1929). Ed. Friedrich-Wilhelm von Herrmann, 1977.
GA 5 *Holzwege* (1935–46). Ed. Friedrich-Wilhelm von Herrmann, 1977.
"Der Spruch des Anaximander" (1946).
"Der Ursprung des Kunstwerkes" (1935–36).
GA 9 *Wegmarken* (1914–70). Ed. Friedrich-Wilhelm von Herrmann, 1976.
"Vom Wesen der Wahrheit" (1930–32).
GA 13 *Aus der Erfahrung des Denkens* (1910–1976). Ed. Hermann Heidegger, 1983.
"Die Kunst und der Raum" (1969).
GA 19 *Platon: Sophistes* (1924–25). Ed. Ingeborg Schußler, 1992.
GA 20 *Prolegomena zur Geschichte des Zeitbegriffs* (1925). Ed. Petra Jaeger, 1988.
GA 24 *Die Grundprobleme der Phänomenologie* (1927). Ed. Friedrich-Wilhelm von Herrmann, 1975.
GA 25 *Phänomenologische Interpretation von Kants Kritik der reinen Vernunft* (1927–28). Ed. Ingtraud Görland, 1987.
GA 26 *Metaphysische Anfangsgründe der Logic im Ausgang von Leibniz* (1928). Ed. Klaus Held, 1978.
GA 34 *Vom Wesen der Wahrheit. Zu Platons Hölengleichnis und Theätet* (1931–32). Ed. Hermann Mörchen, 1988.
GA 40 *Einführung in die Metaphysik* (1935). Ed. Petra Jaeger, 1983.
GA 43 *Nietzsche: "Der Wille zur Macht als Kunst"* (1936–37). Ed. Bern Heinbüchel, 1985.
GA 45 *Grundfragen der Philosophie. Ausgewählte "Probleme" der "Logik"* (1937–38). Ed. Friedrich-Wilhelm von Herrmann, 1984.
GA 53 *Hölderlins Hymne "Der Ister"* (1943). Ed. Walter Biemel, 1984.
GA 54 *Parmenides* (1942–43). Ed. Manfred S. Frings, 1982.
GA 55 *Heraklit, "Der Anfand des Abendländichen Denkens (Heraklit)"* (1943). Ed. Manfred S. Frings, 1979.
GA 65 *Beiträge zur Philosophie (Vom Ereignis)* (1936–38). Ed. Friedrich-Wilhelm von Herrmann, 1989.
GA 66 *Besinnung* (1937–38). Ed. Friedrich-Wilhelm von Herrmann, 1997.
GA 69 *Geschichte des Seins* (1938–40). Ed. Peter Trawny, 1998.
Forthcoming, *Eine Auseinandersetzung mit Sein und Zeit.*

Other Works by Heidegger

"A Dialogue On Language." In *On the Way to Language,* tr. Peter D. Hertz. San Francisco: Harper, 1976.

"Aletheia (Heraklit Fragment 16)." In *Vorträge und Aufsätze.* Stuttgart: Neske, 1985.

"The Anaximander Fragment." In *Early Greek Thinking,* tr. David Farrell Krell and Frank A. Capuzzi. New York: Harper & Row, 1975.

Art and Space. Tr. Charles H. Seibert. *Man and World* 9 (1973): 3–8.

L'arte e lo spazio. Tr. Carlo Angelino, introduction by Gianni Vattimo. Genoa: Il Melangolo, 2000.

Basic Question of Philosophy: Selected "Problems" of "Logic." Tr. Richard Rojcewicz and André Schuwer. Indianapolis: Indiana University Press, 1994.

Basic Writings: From 'Being and Time' (1927) to 'The Task of Thinking' (1964). Ed. David Farrell Krell. San Francisco: Harper, 1993.

The Basic Problems of Phenomenology. Tr. Albert Hofstadter. Indianapolis: Indiana University Press, 1988.

"Bauen, Wohnen, Denken." In *Vorträge und Aufsätze.* Stuttgart: Neske, 1985.

Being and Time. Tr. John Macquarrie and Edward Robinson. New York: Harper, 1962.

Being and Time. Tr. Joan Stambaugh. Albany: SUNY Press, 1996.

Contributions to Philosophy: From Enowning. Tr. Parvis Emad and Kenneth Maly. Indianapolis: Indiana University Press, 1999.

Corpo e spazio. Tr. Francesca Bolino. Genoa: Il Melangolo, 2000.

"Dichterische wohnet der Mensch." In *Vorträge und Aufsätze.* Stuttgart: Neske, 1985.

Early Greek Thinking. Tr. David Farrell Krell and Frank A. Capuzzi. New York: Harper & Row, 1975.

Einführung in die Metaphysik. 5th ed. Tübingen: Niemeyer, 1987.

History of the Concept of Time: Prolegomena. Tr. Theodore Kisiel. Indianapolis: Indiana University Press, 1992.

Kant and the Problem of Metaphysics. Tr. Richard Taft. Indianapolis: Indiana University Press, 1990.

"Nature of Language." In *On the Way to Language,* tr. Peter D. Hertz. San Francisco: Harper, 1971.

Nietzsche, vol. 1–2. Tr. and ed. David Farrell Krell. San Francisco: Harper, 1991.

Nietzsche, vol. 3–4. Tr. Frank A. Capuzzi, David Farrell Krell, and Joan Stambaugh, ed. David Farrell Krell. San Francisco: Harper, 1991.

On Time and Being. Tr. Joan Stambaugh. New York: Harper & Row, 1972.

On the Way to Language. Tr. Peter D. Hertz. San Francisco: Harper, 1971.

"Origin of the Work of Art, The." In *Basic Writings,* ed. David Farrell Krell. San Francisco: Harper, 1993.

"Phänomenologischen Interpretationen zu Aristoteles: Anzeige der Hermeneutischen Situation." In *Dilthey-Jahrbuch für Philosophie und Geschichte der Geisteswissenschaften,* vol. 6. Göttingen: Vandenhoeck & Ruprecht, 1989.

Phenomenological Interpretation of Kant's 'Critique of Pure Reason.' Tr. Parvis Emad and Kenneth Maly. Indianapolis: Indiana University Press, 1997.

"Protokoll zu einem Seminar über den Vortrag 'Zeit und Sein.'" In *Zur Sache des Denkens*, 3d ed. Tübingen: Niemeyer, 1988.

Sein und Zeit. Tübingen: Niemeyer, 1986.

Unterwegs zur Sprache. Pfüllingen: Neske, 1986.

"Vom Ursprung des Kunstwerks. Erste Ausarbeitung (1931–32)." *Heidegger Studies* 5 (1989): 5–22.

"Zeit und Sein." In *Zur Sache des Denkens*, 3d ed. Tübingen: Niemeyer, 1988.

Zollikoner Seminare: Protokolle, Gespräche, Briefe. Frankfurt am Main: V. Klostermann, 1987.

Zur Sache des Denkens, 3d ed. Tübingen: Niemeyer, 1988.

"Zur Überwindung der Aesthetik. Zu 'Ursprung des Kunstwerks' (1934)." *Heidegger Studies* 6 (1990): 5–7.

Other Primary Works

Aristotle. *Aristoteles Physik*, vols. 1–4. Hamburg: Felix Mainer, 1987.

———. *Coming to Be and Passing Away.* Tr. E. S. Forster and J. D. Furley. Cambridge, Mass.: Harvard University Press, 1987.

———. *On Interpretation.* Tr. H. P. Cooke. Loeb Classical Library. Cambridge, Mass.: Harvard University Press, 1969.

———. *Metaphysics.* Tr. Hugh Tredennick. Loeb Classical Library. Cambridge, Mass.: Harvard University Press, 1989.

———. *The Nicomachean Ethics.* Tr. Harris Rackham. Loeb Classical Library. Cambridge, Mass.: Harvard University Press, 1990.

———. *Physics*, vol. 1. Tr. Phillip Wicksteed and Francis Cornford. Loeb Classical Library. Cambridge, Mass.: Harvard University Press, 1963.

———. *Posterior Analytics.* Tr. Hugh Tredennick and E. S. Forster. Loeb Classical Edition. Cambridge, Mass.: Harvard University Press, 1930.

Descartes, René. *The Philosophical Writings of Descartes*, vol. 1. Tr. John Cottingham, Robert Stoothoff, and Dugald Murdoch. Cambridge: Cambridge University Press, 1988.

Kant, Immanuel. *Critique of Pure Reason.* Tr. Norman Kemp Smith. New York: St. Martin's Press, 1965.

———. *Kritik der reinen Vernunft.* Hamburg: Felix Meiner, 1956.

———. *Was Heißt sich im Denken orientieren.* Tr. Petra Dal Santo. Milan: Adelphi, 1996.

Plato. *Phaedo.* Tr. Harold North Fowler. Loeb Classical Library. Cambridge, Mass.: Harvard University Press, 1990.

———. *The Republic*, vols. 1–2. Tr. Paul Shorey. Loeb Classical Library. Cambridge, Mass.: Harvard University Press, 1982.

———. *Sophist.* Tr. Eva Brann, Peter Kalkavage, and Erik Salem. Newburyport: Pullins Company, 1996.

———. *Timaeus.* Tr. R. G. Bury. Loeb Classical Library. Cambridge, Mass.: Harvard University Press, 1989.

Secondary Works

Agamben, Giorgio. *Language and Death: The Place of Negativity.* Tr. Pinkus and Hardt. Minneapolis: University of Minnesota Press, 1991.

———. *Potentialities: Collected Essays on Philosophy.* Tr. Daniel Heller-Roazen. Stanford: Stanford University Press, 1999.

———. *Infanzia e Storia: Distruzione dell'Esperienza e Origine della Storia.* Turin: Einaudi, 1978.

———. *L'ouvert: De l'homme et de l'animal.* Tr. J. Gayraud. Paris: Rivages, 2002.

Beck, Lewis White. *Early German Philosophy: Kant and His Predecessors.* Bristol: Thoemmes Press, 1969.

Becker, Oskar. *Beiträge zur phänomenologischen Begründung der Geometrie und ihrer physikalischen Anwendung,* 2d ed. Tübingen: Niemeyer, 1973.

Blanchot, Maurice. *The Unavowable Community.* Tr. P. Joris. New York: Station Hill Press, 1983.

———. "Michel Foucault as I Imagine Him." In *Foucault, Blanchot.* New York: Zone Books, 1990.

———. *The Writing of the Disaster.* Tr. Ann Smock. Lincoln: University of Nebraska Press, 1995.

Boeder, Heribert. *Seditions: Heidegger and the Limit of Modernity.* Ed. and tr. Marcus Brainard. Indianapolis: Indiana University Press, 1997.

Bonola, Massimo, ed. *Hannah Arendt Martin Heidegger Lettere.* Turin: Edizione di Comunita, 2001.

Brague, Rémi. *Aristote et la Question du Monde.* Paris: Presses Universitaires de France, 1988.

Casey, Edward. *Getting Back Into Place.* Indianapolis: Indiana University Press, 1993.

———. *The Fate of Place.* Berkeley and Los Angeles: University of California Press, 1997.

———. "Smooth Spaces and Rough-Edged Places: The Hidden History of Place." *The Review of Metaphysics* 51 (Dec. 1997): 267–98.

Celan, Paul. *Selected Poems and Prose of Paul Celan.* Tr. John Felstiner. New York: W. W. Norton, 2001.

Chantraine, P. *Dictionnaire Etymologique de la Langue Grecque,* vol. 4. Paris: Klicksieck, 1980.

Cornford, Francis M., ed. and tr. *Plato's Cosmology: The 'Timaeus' of Plato.* London: Routledge, 1952.

Cunliffe, R. *A Lexicon of the Homeric Dialect.* Norman: University of Oklahoma Press, 1988.

Dastur, Françoise. "Heidegger's Freiburg Version of 'The Origin of the Work of Art.'" In *Heidegger Toward the Turn,* ed. James Risser. Albany: SUNY Press, 1999.

De Beistegui, Miguel. *Heidegger and the Political.* New York: Routledge, 1998.

Deleuze, Gilles, and Felix Guattari. *A Thousand Plateaus: Capitalism and Schizophrenia.* Minneapolis: University of Minnesota Press, 1987.

Derrida, Jacques. *Marges de la Philosophie.* Paris: Minuit, 1972.

———. "*Ousia* and *Gramme:* Note on a note from Being and Time." In *Margins of Philosophy,* tr. Alan Bass. Chicago: University of Chicago Press, 1972.

———. *Eperons: les styles de Nietzsche.* Paris: Flammarion, 1974.
———. *De l'esprit: Heidegger et la question.* Paris: Galilée, 1983.
———. "Geschlecht II: Heidegger's Hand." In *Deconstruction and Philosophy,* ed. John Sallis. Chicago: University of Chicago Press, 1987.
———. "Geschlecht I," "Geschlecht II," "De l'esprit." In *De l'esprit: Heidegger et la question.* Paris: Galilée, 1993.
———. *A Derrida Reader: Between the Blinds.* Ed. P. Kamuf. New York: Columbia University Press, 1991.
———. "Geschlecht: Sexual Difference, Ontological Difference." In *A Derrida Reader: Between the Blinds,* ed. P. Kamuf. New York: Columbia University Press, 1991.
———. *Aporias.* Tr. Thomas Dutoit. Stanford: Stanford University Press, 1993.
———. *Khôra.* Paris: Galilée, 1993.
———. *On the Name.* Ed. Thomas Dutoit, tr. David Wood, J. P. Leavey, and I. McLeon. Stanford: Stanford University Press, 1995.
———. *Dissemination.* Tr Barbara Johnson. Chicago: University of Chicago Press, 1981.
Dreyfus, Hubert. *Being-in-the-World.* Cambridge: MIT Press, 1991.
Figal, Günter. *Martin Heidegger Phänomemnologie der Freiheit.* Frankfurt am Main: Athenaum, 1988.
Franck, Didier. *Heidegger et le probème de l'espace.* Paris: Minuit, 1986.
Friedlander, Paul. *Platon: Seinswahrheit und Lebenswerklichkeit.* Berlin: Gruyter & Co., 1954.
———. *Plato: An Introduction,* vol. 1. Tr. Hans Meyerhoff. New York: Bollinger, 1958.
Frisk, H. *Griechisches Etymologisches Wörterbuch,* vol. 2. 2d ed. Heidelberg: Universitätsverlag, 1973.
Fóti, Véronique. *Heidegger and the Poets.* Atlantic Highlands, N.J.: Humanities Press International, 1992.
Foucault, Michel. *Les mots et les choses.* Paris: Gallimard, 1966.
———. *Language, Counter-memory, Practice: Selected Essays and Interviews.* Ed. Donald F. Bouchard. Ithaca: Cornell University Press, 1977.
———. *Power/Knowledge: Selected Interviews and Other Writings 1972–1977.* Ed. Colin Gordon. New York: Pantheon Books, 1980.
———. *Dits et Ecrits: 1954–1988,* vols. 1–4. Paris: Gallimard, 1989.
———. "Des espace autres." In *Dits et Ecrits.*
———. "Espace, savoir [. . .]." In *Dits et Ecrits.*
———. "Le langage de l'espace." In *Dits et Ecrits.*
———. "Maurice Blanchot: The Thought from Outside." In *Foucault, Blanchot.* New York: Zone Books, 1990.
———. *Politics, Philosophy, Culture: Interviews and other Writings, 1977–1984.* Ed. Lawrence D. Kritzman. New York: Routledge, 1998.
Gadamer, Hans-Georg. "Plato." In *Heidegger's Ways,* tr. John W. Stanley. Albany: SUNY Press, 1994.
Gasché, Rodolphe. *The Tain of the Mirror.* Cambridge, Mass.: Harvard University Press, 1986.
Haar, Michel. *Le Chant de la Terre: Heidegger et les assises de l'histoire de l'être.* Paris: L'Herne, 1985.

———. "La relation monde-terre chez Heidegger." In *L'oeuvre d'art.* Paris: Hartier, 1994.

Herrmann, Friedrich-Wilhelm von. *Heideggers Philosophie der Kunst: Eine systematische Interpretation der Holzwege-Abhandlung 'Der Ursprung des Kunstwerkes.'* Frankfurt am Main: V. Klostermann, 1980.

———. *Subjekt und Dasein.* Frankfurt am Main: V. Klostermann, 1985.

———. "Von 'Sein und Zeit' zum 'Ereignis.'" In *Von Heidegger Her,* ed. Hans-Helmuth Gander. Frankfurt am Main: V. Klostermann, 1991.

———. *Wege ins Ereignis.* Frankfurt am Main: V. Klostermann, 1994.

———. "Wahrheit-Zeit-Raum." In *Die Frage nach der Wahrheit,* ed. Ewald Richter. Frankfurt am Main: V. Klostermann, 1997.

Hesiod. *The Homeric Hymns and Homerica.* Tr. Hugh Evelyn-White. Loeb Classical Library. Cambridge, Mass.: Harvard University Press, 1936.

Historisches Wörterbuch der Philosophie, vol. 8. Ed. Joachim Ritter and Karlfried Gründer. Darmstadt: Wissenschaftliche Buchgesellschaft, 1992.

Hölderlin, Friedrich. *Poems and Fragments.* Tr. Michel Hamburger. New York: Cambridge University Press, 1980.

Husserl, Edmund. "Husserl On Space and Time." In *Husserl: Shorter Works,* ed. Peter McCormick and Frederick A. Elliston. Notre Dame: University of Notre Dame Press, 1981.

Irigaray, Luce. *The Forgetting of Air in Martin Heidegger.* Tr. Mary Beth Mader. Austin: University of Texas Press, 1983.

Jähnig, Dieter. "Die Kunst und der Raum." In *Erinnerung an Martin Heidegger,* ed. Günther Neske. Pfüllingen: Neske, 1977.

Kettering, Emil. *Nähe: Das Denken Martin Heideggers.* Pfüllingen: Neske, 1987.

King, Magda. *A Guide to Heidegger's 'Being and Time.'* Ed. John Llewellyn. Albany: SUNY Press, 2001.

Kissiel, Theodore. *The Genesis of Heidegger's Being and Time.* Berkeley and Los Angeles: University of California Press, 1993.

Klein, Jacob. *A Commentary On Plato's 'Meno.'* Chicago: University of Chicago Press, 1965.

———. *J. Klein: Lectures and Essays.* Annapolis: St. John's College Press, 1985.

———. *Greek Mathematical Thought and the Origin of Algebra.* Tr. Eva Brann. New York: Dover, 1992.

Krell, David Farrell. "Being and Time." In *Basic Writings: From 'Being and Time' (1927) to 'The Task of Thinking' (1964): Martin Heidegger.* New York: Harper & Row, 1977.

Kristeva, Julia. *Étrangers à nous-mêmes.* Paris: Gallimard, 1988.

———. *Strangers to Ourselves.* Tr. Leon Roudiez. New York: Columbia University Press, 1991.

Lévêque, Pierre, and Pierre Vidal-Naquet. *Cleisthenes the Athenian.* Tr. David Ames Curtis. Atlantic Highlands, N.J.: Humanities Press, 1996.

Liddell, Henry George, and Robert Scott. *A Greek-English Lexicon.* Oxford: Oxford University Press, 1989.

Ludz, Ursula, ed. *Briefe 1925–1975: Und andere Zeugnisse* (Frankfurt am Main: V. Klostermann, 1998).

Lyotard, Jean-François. *Heidegger and 'the Jews.'* Tr. Andreas Michel and Mark S. Roberts. Minneapolis: University of Minnesota Press, 1995.

Macann, Christopher, ed. *Martin Heidegger: Critical Assessments.* London: Routledge, 1992.

Malpas, J. *Place and Experience: A Philosophical Topography.* Cambridge: Cambridge University Press, 1999.

Marx, Werner. *Heidegger and the Tradition.* Tr. Theodore Kissiel and Murray Greene. Evanston: Northwestern University Press, 1971.

Merleau-Ponty, Maurice. *Phenomenology of Perception.* Tr. Colin Smith. New York: Routledge, 1994.

Neruda, Pablo. *Veinte Poemas de Amor . . . Cien Sonetos de Amor.* Barcelona: Plaze y James, 1996.

Oxford Classical Dictionary. Ed. Hornblower and Spawforth. Oxford: Oxford University Press, 1996.

Patocka, Jan. *Qu'est-ce que La Phenomenologie?* Tr. from German by Erika Abrams. Grenoble: Jerome Millon, 1988.

Poggeler, Otto. *Der Denkweg Martin Heideggers.* Pfüllingen: Neske, 1983.

———. *Die Frage nach der Kunst. Von Hegel zu Heidegger.* Freiburg: Munchen, 1984

———. "Heidegger und die Kunst." In *Martin Heidegger: Kunst, Politik, Technik* Ed. Jamme and Harries. Frieburg: München, 1992.

———. *Über die moderne Kunst: Heidegger und Klee's Jenaer Rede von 1924.* Jena: Erlangen, 1995.

Raffoul, Francois. *Heidegger and the Subject.* Tr. David Pettigrew and Gregory Recco. Atlantic Highlands, N.J.: Humanities Press, 1998.

Ross, William David. *Aristotle.* New York: Routledge, 1995.

Sallis, John. *Double Truth.* Albany: SUNY Press, 1990.

———. *Echoes: After Heidegger.* Indianapolis: Indiana University Press, 1990.

———. *Reading Heidegger: Commemorations.* Indianapolis: Indiana University Press, 1993.

———. *Stone.* Indianapolis: Indiana University Press, 1994.

———. "Timaeus' Discourse on the Chora." Lecture: Boston University, 1995.

———. *Being and Logos.* Indianapolis: Indiana University Press, 1996.

———. *Shades—Of Painting at the Limit.* Indianapolis: Indiana University Press, 1998.

———. *Chorology: On Being In Plato's 'Timaeus.'* Indianapolis: Indiana University Press, 1999.

Schmidt, Dennis. *The Ubiquity of Infinity: Hegel, Heidegger, and the Entitlement of Philosophy.* Cambridge: MIT Press, 1988.

———. *On Germans and Other Greeks: Tragedy and Ethical Life.* Indianapolis: Indiana University Press, 2002.

Scott, Charles E. *The Language of Difference.* Atlantic Highlands, N.J.: Humanities Press, 1987.

———. *The Question of Ethics.* Indianapolis: Indiana University Press, 1990.

———. "Heidegger, Madness, and Well-being." In *Martin Heidegger: Critical Assessments,* ed. Chris Macann. New York: Routledge, 1992.

———. "*Adikia* and Catastrophe: Heidegger's 'Anaximander Fragment.'" *Heidegger Studies* 10 (1994) :127–42.
———. *On the Advantages and Disadvantages of Ethics and Politics.* Indianapolis: Indiana University Press, 1996.
———. *The Time of Memory.* Albany: SUNY Press, 1999.
Scott, Charles E., Susan Schoenbohm, Daniela Vallega-Neu, and Alejandro Vallega, eds. *A Companion to Heidegger's 'Contributions to Philosophy.'* Indianapolis: Indiana University Press, 2001.
Simplicius. *Commentaire Sur les Categories d'Aristote,* vol. 2. Tr. Guillaume de Moerbeke. Leiden: E. J. Brill, 1975.
Socrates. *Critias.* Tr. R. G. Bury. Loeb Classical Library. Cambridge, Mass.: Harvard University Press, 1989.
———. *Phaedo.* Ed. C. J. Rowe. Cambridge Greek and Latin Classics. Cambridge: Cambridge University Press, 1993.
Sophocles. *Antigone.* In *Greek Tragedies,* vol. 1, ed. David Green and Richmond Lattimore. Chicago: University of Chicago Press, 1991.
Steinbock, Anthony J. *Home and Beyond: Generative Phenomenology after Husserl.* Evanston: Northwestern University Press, 1995.
Taylor, Alfred E. *A Commentary On Plato's 'Timaeus.'* Oxford: Clarendon, 1972.
Thompson, John Eric Sidney. *Maya History and Religion.* Norman: University of Oklahoma Press, 1979.
Vallega-Neu, Daniela. *Die Notwendigkeit der Gründung im Zeitalter der Dekonstrution. Zur Gründung in Heideggers "Beiträgen zur Philosophie" unter Hinzuziehung der Derridaschen Dekonstruktion.* Berlin: Duncker & Humblot, 1997.
———. "Abysmal Grounding in Heidegger's 'Contributions to Philosophy.'" Proceeding, Heidegger Conference of America, 1997.
———. *Heidegger's Contributions to Philosophy: An Introduction.* Bloomington: Indiana University Press, 2003.
Villa-Petit, M. "Heidegger's Conception of Space." In *Martin Heidegger: Critical Assessments,* ed. Chris Macann. London: Routledge, 1992.

Index